Decolonizing Geography

Decolonizing the Curriculum

Ali Meghji, *Decolonizing Sociology*
Robbie Shilliam, *Decolonizing Politics*
Sarah A. Radcliffe, *Decolonizing Geography*

Decolonizing Geography

An Introduction

Sarah A. Radcliffe

polity

First published in 2022 by Polity Press

Polity Press
65 Bridge Street
Cambridge CB2 1UR, UK

Polity Press
101 Station Landing
Suite 300
Medford, MA 02155, USA

ISBN-13: 978-1-5095-4159-1
ISBN-13: 978-1-5095-4160-7(pb)

A catalogue record for this book is available from the British Library.

Library of Congress Control Number: 2021945136

Typeset in 10.5 on 12.5pt Sabon
by Fakenham Prepress Solutions, Fakenham, Norfolk NR21 8NL
Printed and bound in Great Britain by TJ Books Ltd, Padstow, Cornwall

The publisher has used its best endeavours to ensure that the URLs for external websites referred to in this book are correct and active at the time of going to press. However, the publisher has no responsibility for the websites and can make no guarantee that a site will remain live or that the content is or will remain appropriate.

Every effort has been made to trace all copyright holders, but if any have been overlooked the publisher will be pleased to include any necessary credits in any subsequent reprint or edition.

For further information on Polity, visit our website:
politybooks.com

For a World Where Many Worlds Fit

Contents

Author's Note

For clarity, key terms appear in bold when first introduced. These terms are explained in the Glossary section. Emboldened references in brackets refer to sections, chapters, figures or textboxes in this book.

Preface

As with any piece of academic writing, this book reflects its author and where she stands in the context of social, institutional and geopolitical relations. As such, *Decolonizing Geography: An Introduction* is deeply situated and is not about decolonizing everywhere. It emerges principally out of Anglophone postcolonial and decolonial geography and Anglophone geographers' critical engagements with numerous Other geographies and knowledges around the world. As such, the book speaks back to the global predominance of Anglophone geography in former colonial and settler colonial countries where racialization, the westernizing university and settler colonialism operate and are challenged. Brazilian, Mexican, French and Hungarian geographies, to name a few, have different stories to tell. I encourage all readers to think about this book in tandem with the local and regional decolonizing discussions where they live and work.

My position in these geopolitical and intersectional configurations is as a white, cis-gender woman with an Anglo name in an overwhelmingly white British department of geography. My training and experience are in human geography; the department includes human and physical geographers, the vast majority white, especially among faculty. Geographers of colour have argued rightly that geography's urgent task of decolonizing must not rest solely on racialized minorities. I concur wholeheartedly, and as a white ally stress the importance of white geographers' informing themselves about decolonizing and anti-racism. The construction of a decolonial pluri-geo-graphy – or a world of many worlds – depends on all of us. Plural decolonizing geographies crucially require white geographers to take responsibility for and actively work to overturn racialized exclusions and assumptions. The knowledge geopolitics behind this book additionally reflect my decades of ethnographic work with Latin American scholars, activists and communities, especially in Andean rural districts and with Indigenous groups, leaders and organizations. It is their critiques, experiences of racism and exclusion, and hopeful agendas for change that enliven this book. In terms of its focus, however, the book is written to be accessible and relevant for physical as much as human geographers. The chapters include physical and human geography examples, discussions, and pointers to further reading. The book was also influenced by events during the Covid-19 pandemic which provided daily reminders of coloniality's persistence and of decolonizing ripostes such as the Black Lives Matter movement.

The book aims to broaden understanding of why decolonizing matters among instructors and students in geography and cognate disciplines. Chapters 1 through 4 provide a general introduction addressed particularly to geographers who, like me, are located in westernizing, white-dominated and/or wealthier countries. Chapter 5 deals with issues of teaching and learning, while Chapter 6 covers research of various kinds, including short student projects. To make the decolonizing framework and approach more accessible, a Glossary at the end of the book provides definitions of terms used in the book. North American, European and Australasian geographies appear throughout, although their tertiary education systems and terminologies vary. I have tried to avoid too many British-isms! Across these regions, geographers differ in whether and how they self-identify in racial-ethnic and territorial terms; I provide this information where available but cannot do so consistently. This book addresses exciting and rapidly moving debates which shift as activism and scholarship consider important dimensions related to colonialism. This context emphasizes the urgency for geography and geographers to change their approaches, materially and on short time scales. So, while reading this book, I encourage readers to put it into conversation with blogs, non-academic writings, activism and news stories that speak to decolonizing issues where *you* stand. Finally, in an introductory textbook it was inappropriate to address structural issues connected to neoliberal colonial academia that systematically influence hiring decisions, promotions, funding streams for research and the colonial biases of journals and peer review. These are crucial issues rightly critiqued in other forums.

To acknowledge the support, encouragement and care that made this book possible, I end with some thanks. Thanks to Pascal Porcheron, Stephanie Homer and Ellen MacDonald-Kramer at Polity, who encouraged and cajoled this manuscript to the end, in the nicest ways. I have tried to unlearn ingrained assumptions, so I'm extremely grateful to everyone who pulls me up on partial understandings and privileged blind spots. Key among those who did that are three anonymous readers. Incorporating their suggestions, together with bibliographies and insights into unfamiliar contexts, the book aims to do justice to those plural realities, albeit humbly and provisionally. Friends and colleagues near and far inspired me with writing, action and conversation during the book's conception and writing: a big thank you to Laurie Denyer Willis, Rogerio Haesbaert, Humeira Iqtidar, Anna Laing, Sian Lazar, Monica Moreno Figueroa, Kamal Munir, Nancy Postero, Isabella Radhuber, Catherine Souch, Natasha Tanna, Yvonne Underhill-Sem, Georgie Wemyss and Sofia Zaragocín. I am very grateful to Nicola J. Thomas and Ian Cook for sharing teaching materials and reading a draft chapter. Debates at the Decolonial Research Lab sharpened my thinking; *gracias* to Tiffany Dang, Ellen Gordon, Ana Guasco, Sam Halvorsen, Laura Loyola-Hernandez, Tami Okamoto, Sandra Rodriguez Castañeda and Giulia Torino. Current postgraduate students Matipa Mukondiwa, Emiliano Cabrera Rocha, Ashley Masing and Lily Rubino bring news and plural perspectives to my attention over Zoom. Over the years, final-year students on the geographies of postcolonialism and decoloniality course have prompted me with questions; I hope this text does them justice. The Decolonizing Cambridge Geography working group – especially Sophie Thorpe, Sophia Georgescu, Fran Rigg, Joseph Martinez-Salinas, Josie Chambers, Ollie Banks, Charlotte Millbank, Ed Kiely and Fleur Nash – devised an agenda for departmental change where I work. Taking that agenda to the next level would not have been possible without steady support from Bhaskar Vira, Harriet Allen, Charlotte Lemanski, Michael Bravo, Sam Saville and Phil Howell, among others. Over longer stretches of time and distance, the experiences and voices of Ecuadorian Kichwa *warmikuna* and Tsáchila *sonala* continue to resonate through my thinking and acting on decoloniality; for that, I honour their strength in facing down numerous hurdles, and appreciate their generosity in dialoguing with me. And for my whole family, including a 2021 baby, a thousand thanks for many thousands of moments of love and care.

Sarah A. Radcliffe
Cambridge, September 2021

Foreword: Decolonizing in a North–South Dialogue

Rogério Haesbaert

Decolonizing Geography is a book about action and doing, as all geography books should be; it is essential to look at space through the actions of different actors-subjects, human and more-than-human, in their multiple relations to time and space. Living, indeed, means transforming space and transforming ourselves through space, since it constitutes us in the first place as bodies (or body-territories, as we have learned from Indigenous peoples and Latin American feminists). Consistent with decolonial approaches, our aim should be not only to treat every theoretical approach analytically, but to treat categories of analysis also *in dialogue with* categories of practice – that is, ultimately deriving from common sense and struggles 'from below'. Additionally, these categories are normative in pointing to a new geographic horizon for the future.

In making a decolonizing geography, Sarah Radcliffe has engaged openly in dialogue with what sometimes, in a simplified way, we see as 'the South', as if a well-defined geography was delineated between a North and a South – the North always positioned 'on the top' of the map or compass. Making geography is always about understanding and practising one fundamental characteristic of space in motion, namely its ability to change one's perspective and thereby discover other worlds. Thus, practising space – doing geography – means, above all, seeking to look at the world from the point of view of Others. The book does this masterfully, based on Radcliffe's longstanding and generous life's work alongside peoples and cultures often labelled 'peripheral' (such as Kichwa peoples in Ecuador), and her teaching and learning with them. Indigenous peoples show us today how relative the categories of North, South, centre and periphery are. To decolonize is precisely to have the ability to understand/recognize the Other's gaze and transform ourselves with it, changing our perspective and 'classificatory' vision. Today, indeed, peripheral, Southern and colonized groups bring fundamental lessons that many central or Northern geographers, in their anthropocentric and dominating/classifying zeal, ignored or despised for a long time.

Taking up points emphasized by the author, I would like to focus on the critiques of decolonial approaches, which defenders of this way of *sentipensar* and acting constantly face. A Spanish and Portuguese term used by Latin American decolonial thinkers, *sentipensar* is a neologism that breaks the binary of feeling ('sentir' to feel) and thinking ('pensar' to think). In her book, Sarah Radcliffe warns us not to romanticize pre-colonial societies. These societies were already very complex and differentiated; for instance, some pre-Colombian states such as the Aztecs and Incas practised forms of colonialism with the ambition of dominating and imposing ideologies (albeit far from present-day capitalism's extent and intensity). On the other hand, we must always be attentive to the risk of oversimplifying decolonial critiques of 'modernity', which, despite all its processes of domination, was also the cradle of autonomous thought. The philosopher Cornelius Castoriadis, for instance, says that modernity is based on a constant dispute between two social projects, one heteronomic (domination/subordination) and the other autonomous-liberatory, with the triumph of the former. Likewise, not all ('modern') European thought is Eurocentric, defending the imposition of a modern-Euro-colonial 'one-world world' universalism. In the Latin American case, the situation is even more complex, as North America imposes itself through colonizing power, starting with the name: 'American' designates a resident of the United

States as well as an inhabitant of the entire continent, aspects that reflect the Monroe Doctrine (1823) and its ambiguous motto 'America for [North] Americans'. For this reason, Indigenous peoples in the continent decolonize America by re-naming it Abya Yala ('living earth').

Another dilemma of decolonial thinking is the risk of overemphasizing oppressions of race-ethnicity and gender and downplaying their intersectionality with class domination. Treating all these dimensions as mutually constitutive and contextualizing them geo-historically, however, is no easy task. The designation modernity-coloniality has always been closely linked to capitalism, as for the colonization process, as this book reminds us, can never be dissociated from the expansive impetus of capitalist accumulation and consumption, as exemplified by Latin America's current subjection to the extractive economic model. Thus, the concept of coloniality can never be dissociated from a critical reading of the capitalist world system as a whole.

Sarah Radcliffe also points out that decolonial thinking and attitudes are not new, and link back to the work of several geographers who were concerned with a critical reading 'from below' based on specific spaces and subalternized groups. They established more egalitarian relationships, with the purpose of making geography across North and South. 'Collaborative and Southern geographies have existed for decades, even if they were not always labelled decolonial', says the author. Achieving this more egalitarian North–South relationship is difficult, however, due to the coloniality of language. From my perspective looking from the South, language appears crucial, and northern intellectuals may not appreciate the importance of mastering a foreign language to carry out a decolonization process, fleeing from the (often implicit) belief that what is 'recognized' or what is 'better' is already (or will soon be) published in the hegemonic language of English. During my time as a post-doctoral researcher at the Open University, I was surprised during a seminar when Doreen Massey introduced me and pointed out that English was my fourth language. Only later did I realize the importance of this, as few geography professors spoke a second language, let alone more, whereas in Brazilian universities managing two foreign languages is a necessary condition to pursue a doctorate. Arguably the 'universalizing' character of English today, especially via the internet, significantly accelerates and facilitates communication, but every self-respecting decolonial study necessarily needs a greater involvement with multiple languages, in order to appreciate the worldviews and geographies of subalternized groups.

Among the lessons to be learnt in a 'North–South' dialogue with Latin America is 'anthropophagy' (as the Brazilian writer Oswald de Andrade discussed in his 1928 'Manifesto Antropofágico') – to receive the Other and somehow 'swallow' it and make something else of it. This Latin American hybridity or 'transculturation' (a term from the Cuban essayist Fernando Ortiz) took place in large part, of course, under the violence of colonization. But much hybridity arises from the longstanding societies and politics of original peoples who, even when forcibly transformed, bring forward decolonial proposals such as the one that opens this book, namely to build 'a world where many worlds fit' (*un mundo donde quepan muchos mundos*, to quote the Zapatista movement). Transculturation thus allows the delineation and building of what Radcliffe calls 'decolonial pluri-geo-graphies'. Having more than one world means accommodating non-hierarchical, diverse, worlds (a pluriverse) and overcoming divisions such as between First, Second and Third worlds.

In addition to the geographical concepts discussed in Chapter 4 there is the concept of region (or regionalization processes), which carries strong Eurocentric overtones, as in the world's division into continents. By contrast, in Brazil, Josué de Castro brings a pioneering approach for a 'regionalization from below' (by identifying the regionalizing of Brazil's starving populations), focusing directly on subalternized groups. Mariátegui, the Peruvian Marxist thinker, in turn, speaks of a 'new regionalism' in Peru, centred around Indigenous peoples and land issues. The concept of territory similarly can be decolonized further. The concept of territory in Latin America informs critical geographical accounts because of the term's use in struggles 'not only for land but also for territory', as Indigenous peoples say. As Radcliffe points out, unlike Anglophone geographies' functional and 'technological' definition of territory, here territory is understood as a defensive and *affirmative* space of life, struggling for existence or, as Carlos Walter Porto-Gonçalves expresses it, a *r-existencia* (resisting in order to exist). Always in movement, territory must be seen in the multiplicity of its manifestations and overlaps, in short, as a multi- or trans-territoriality – as Guaraní peoples on the border between Paraguay, Bolivia, Brazil and Argentina claim. This existential 'life-territory' is evident also in Arturo Escobar's discussion of Afro-descendant peoples in Colombia's Pacific region. These geographies alert us to the practical and political relevance of our concepts. As Radcliffe demonstrates here, Latin American Indigenous feminists engage in political activism using the concept of body-territory, which links their 'domain and appropriation' over space to their own bodies (as Sofía Zaragocín indicates when proposing that the female uterus itself generates territoriality).

The 'novelty' of decolonial approaches is therefore not so new if we situate it in relation to diverse Indigenous and Latin American thought. Likewise it is important not to make the so-called decolonial turn into a theoretical paradigm that will impose itself with full force against other ways of thinking about space and doing geography. As Doreen Massey said, we must be very careful because tomorrow 'our own theory' will be questioned and surpassed. Hence decolonizing entails overcoming the idea of radical paradigm shifts and instead promotes coexistence between diverse approaches. As Radcliffe states: 'to ensure geography transforms into a discipline appropriate for a world "where many worlds fit", this analytical plurality is crucial. Indeed, acknowledging *plural* theoretical reference points is entirely fitting, being consistent with decolonial agendas to acknowledge and value multiple systems of knowledge.'

Finally, this book calls on everyone to carry out their own plural decolonizations from the spatiotemporal and geo-historical contexts in which they are situated. 'I encourage all readers to think about this book in tandem with the local and regional decolonizing discussions where they live and work.' Clearly recognizing the situation in which our knowledge is conceived is the first indispensable step for the construction of decolonizing dialogues with human and more-than-human Others, dialogues that expand our views of the world(s). Decolonizing is, ultimately, about proposing the challenge of new ways of building power relations, of making politics that is always spatiotemporally situated, attentive to the limits of the act of (dis)ordering space (including concerns about the rights of nature), and to all kinds of inequalities and/or differences.

In summary, this book can be read not only for its analytical vigour and innovative approach to space and geography, but also as a stimulus for action. In times as difficult as these in which we live, especially for subalternized populations in the majority world 'periphery', this book conveys encouragement as well as critique, dialogue as well as action. Decolonizing geography, in Sarah Radcliffe's book, recognizes that there are many legitimate ways of reading and making space, and that our greatest struggle and challenge is to embrace this diversity of world perspectives while tackling its inequality.

Figures, Tables and Boxes

–1–
Why Decolonize Geography?

I The Geographies of Coloniality

The contemporary world witnesses relations of power, organization of space, priorities, and mindsets that are deeply influenced by patterns of hierarchy and domination that originated in modern forms of **colonialism** and empire. While not unchanged, these colonial relations, organizations and mindsets are present here and now, from Australia to Canada, Mexico to Finland. We call these patterns of power **coloniality**, a term that alerts us to the systematic exclusions and narrow interpretations that define early twenty-first-century modernity. Against this background, the process of **decolonizing** offers a multifaceted programme to identify and challenge coloniality's material, institutional and ideological outcomes. Coloniality and decolonizing are intertwined dimensions of the modern world, found all around us in urban landscapes, universities, political arrangements, and ideas about nature.

This book provides an introduction to decolonizing geography. Geography here refers to the world in all its variation, as well as to the academic discipline that researches and teaches about that world. Decolonizing geography involves mapping configurations of coloniality as they

> ## Coloniality
>
> Coloniality refers to mindsets, knowledges, identities and structures of power that have persisted over centuries. Coloniality became the modern globally dominant socio-spatial system from the late fifteenth century onwards. It comprises dynamic economic, political, social and cultural processes, which combine in various ways across time and space (see **section 1.II**).

touch down in a place, making it unique and connecting it with stretched out spatial structures, flows and mentalities. Coloniality is not the sole influence on geographies, but it has been largely taken for granted. Geographical scholarship has largely overlooked it because the discipline itself is steeped in coloniality. Many commentators argue that geography has a very long way to go before it can contribute to decolonizing – they are right, for reasons this chapter will explain. Yet the process of decolonizing geography's communities and analytical lenses, the book suggests, has the potential to generate important insights into coloniality's operations across space and the shaping of geographical imaginations and theorizations; indeed, in principle the discipline contains vital practices and agendas for real decolonizing change. This introductory chapter provides an entry point into understanding what coloniality is, why geographers should take it into account, and what decolonizing aims to achieve. We start with a case study, which demonstrates the operations of coloniality in a specific place. This, like the many other physical and human geography examples throughout the book, will show how coloniality operates and why and how decolonizing actions and thought challenge it.

Coloniality's urban landscapes and decolonizing action

In July 2020, a black resin statue of Bristol resident Jen Reid appeared on a plinth where, until a month previously, a bronze figure of the slave-trader Edward Colston had stood for over a century (**Figure 1.1**). Prior to the appearance of the Jen Reid statue, protesters against the racist murder of George Floyd in Minneapolis had gathered in central Bristol in support of the Black Lives Matter movement. During these protests, the Colston statue was taken from its central city location and tipped into the harbour.[1] Now ringed by art galleries, museums and new housing, the harbour had served for centuries as a major hub in the trans-Atlantic slave trade. Companies led by men like Colston had profited from the **enslavement** of Black Africans and from the international trade in human labour, tobacco, sugar and tropical fruit.

Edward Colston was a seventeenth-century Bristol merchant working for the Royal Africa Company, whose monopoly on West African trade in gold, slaves and ivory began in 1662. Between the sixteenth century and the nineteenth, British companies and traders like Colston controlled the mercantilist and then capitalist economic system, which accumulated bullion through trade and enslavement. Like Spanish colonialism in South America, the British traders' pursuit of gold and silver was based on the enslavement of Africans. It is estimated that around half a million enslaved Africans were moved in Bristol ships over the period 1698–1807 (Bristol Black Archives Partnership undated: 3). In the age of mercantilism and imperial **capitalism** in the nineteenth century, only a small number of enslaved and free Africans (and later Afro-Caribbeans) lived in Bristol, but that situation changed in the 1940s.

After the Second World War, the British government invited people from what were then its colonies and dominions, including the Caribbean colonies of Jamaica, Barbados, St Kitts, Nevis and Dominica, to take up jobs in Britain. In the memorable phrase of the British-Sri Lankan intellectual Ambalavaner Sivanandan (1923–2018), 'we [migrants from the former colonies, dominions and protectorates] are here [in the UK] because you [British colonizers] were there'. Today,

---

[1] A video of the toppling of the Colston statue is available at https://www.dropbox.com/s/76tkoqt7zzm3jh5/Colston%20Falls%20Movie%20final%20cut.mp4?dl=0.

Imperialism

Imperialism is a type of geopolitical relation whereby one state ('empire') dominates the political authority of another state or territory through formal (including military and administrative) or informal (cultural, economic) means. Imperialism is a broad category that includes US post-war influence, direct colonization and diverse forms of colonialism.

Bristol is multiracial, reflecting colonial-modern economic structures, geopolitical ties across the world, migrant flows, and – underlying them in turn – a mindset that assumes certain groups and places serve the interests of a former colonial country. Colonial-modern relations thus very much exist in the present day, linking historic relations between places and people to entrenched patterns that embed themselves firmly in cities, interpersonal relations and the prevailing 'common sense'. The British slave trade was ended in the early nineteenth century, and Bristol's economy is now based on the aviation industry, tourism, media, information technology and financial services. Yet urban inequalities for Black residents are longstanding, and exist alongside the injustice of the city's commemorations of colonialism and enslavement.

Looking closely at coloniality raises critical questions about how Colston is lauded as a philanthropist, while Black residents of Bristol face stark inequalities and **racism**, with negative impacts on housing, jobs and policing. Focusing on colonial-modern economy, power and society also sheds light on the reasons behind the toppling of Colston's statue in protests against the racist murder of George Floyd thousands of miles away. In further protests against police brutality in both the USA and the UK, the colonial structure of racism was also at issue. The two Bristol statues – of Jen Reid and Edward Colston – exemplify the relations of power in what the geographer Derek Gregory terms the 'colonial present'. The colonial present consists, he argues, in the Anglo-American amnesia towards colonial pasts and a nostalgia for the British Empire, expressed through interventions to make **Other** (non-Anglo-American) people's geographies serve their own purposes. The colonial present informed the military intervention in the Middle East after 9/11 (Gregory 2004), but it also applies to Bristol's situation and elsewhere. Analysis of the Bristol statues' contexts and meanings reveals the spatially situated workings of coloniality and of the struggles against it – key themes in decolonizing geography.

Decolonizing

Decolonizing refers to practices and processes that actively seek to delink from coloniality. Decolonizing comprises 'a long-term process involving the bureaucratic, cultural, linguistic and psychological divesting of colonial power' (Tuhiwai Smith 2012: 33).

Decolonizing action draws attention to the most enduring and exclusionary dynamics of power at work in the world today, and seeks ways to undo them. Precisely due to its shape-shifting over time and space, coloniality is not unchanging; it has been – and will continue to be – challenged and contested. Critical **anti-racism** and decolonial voices make visible enslavement's consequences in unequal job opportunities, under-resourced neighbourhoods, and narratives of imperial greatness that ignore everyday racism. Bringing coloniality to light refuses complicity with it, and seeks to reorient institutions, practices and frameworks of understanding that fuel and legitimize coloniality. In this respect, decolonizing offers hopeful agendas for transformed futures (**section 1.VI**). In Bristol, after the toppling of the Colston statue, the mayor Marvin Rees called for action to 'make the legacy of today about the future of our city, tackling racism and inequality. I call on everyone to challenge racism and inequality in every corner of our city, and wherever we see it.'[2]

---

[2] Rees's full statement can be found at https://news.bristol.gov.uk/news/statement-from-the-mayor-of-bristol.

As coloniality touches down in a place, so decolonizing futures are envisioned and constructed, resulting in diverse and vibrant action and thinking. As the Bristol case exemplifies, although colonialism is often imagined to be 'over there', coloniality and decolonizing are present here and now. Decolonizing is very much about geography, just as geography has a lot to say about decolonizing, as later chapters will show.

Entitled 'A Surge of Power', the statue of protestor Jen Reid (**Figure 1.1**) recreates and commemorates the moment she leapt onto the recently emptied Colston plinth. Colston's statue had become an issue of concern for many Bristol residents. From the 1990s onwards, campaigners argued that the statue celebrated an oppressive and dehumanizing industry and petitioned for its removal. They challenged historical narratives that place slavery solely in 'the past', and drew attention to ongoing patterns of exclusion in Bristol and beyond. On Anti-Slavery Day in 2018, artist-activists placed figures identified with modern-day slavery jobs, including nail-bar staff and agricultural labourers, next to the statue, in an example of 'guerrilla memorialisation' (Rice 2012). Critical questions were asked about why Colston's statue was erected some 175 years after his death (at the height of the British Empire), and declared a heritage monument in 1977 when the city's Society of Merchant Venturers were actively involved in his commemoration. These decisions to honour rather than question Colston reflect colonial mindsets and **knowledge**. Decolonizing, by contrast, asks critical questions in order to understand coloniality in a given location in relation to wider colonial-modern sites and flows. Decolonial thinking and doing – 'praxis' – thereby seek to undo coloniality in an institution or place and to challenge relations of domination and exclusion.

The remainder of this chapter provides an introductory guide to the operation of coloniality (section II), and an overview of forms of colonialism and the messy outcomes of independence for former colonies (section III). There then follows a discussion of two key pillars of coloniality: the modes of thought and power behind its enduring influence

Figure 1.1 'A Surge of Power (Jen Reid)', statue by Marc Quinn
Source: Wikimedia Commons

(section IV), and the structured exclusions arising from racial hierarchies (section V). In each case, the geography discipline's problematic connections with coloniality are examined, leading to a discussion of the reasons for and the routes to decolonizing geography (section VI), demonstrated through an extended case study of decolonizing Los Angeles. The chapter ends by making the case for decolonizing geography, from physical and environmental geographies to human geography (section VII). A chapter summary and outline of the book follows, together with a list of further reading and resources.

II Coloniality (How to Recognize It) and Decolonizing

The events in Bristol illustrate coloniality in a specific social, urban, political and ideological context. In one sense, Bristol is unique in relation to the characteristics and dynamics of colonialism and decolonizing action. Yet, as this section explains, the city cannot be understood outside the frames of global linkages, enforced racial hierarchies and enduring inequalities that render coloniality such an extensive and enduring influence on the world. Although it exists every-where, coloniality is not expressed uniformly in identical urban landscapes, racial exclusions, and contests over place. Coloniality encompasses diverse economic, social, political and ideological processes that touch down in a place, making it both context-specific *and* interconnected with other spaces and scales. As the Bristol case demonstrates, coloniality is not limited to a colonial or imperial past; its operations are very much in the present.

To address coloniality through action and critical awareness thus requires understanding its operations and concrete dimensions. Coloniality works via the structures, institutions, flows and everyday processes that actively construct systems of power, thinking and behaviour across differentiated social sectors and areas. Identification of key economic, social, political and ideological colonial processes provides the basis for under-standing contemporary expressions of **coloniality-modernity** in its uneven and differentiated pathways across the world. Colonial projects are not external to modernity, but neither do they fully determine social and material relations today, due to both continuities *and* discontinuities, reorganization *and* resistance. The concept of coloniality thus recognizes there is no direct equivalence between the past and the present, yet it does suggest that the key drivers of *connections* between the (colonial-imperial) past and the present can be identified, understood and lead to socio-political change (Ahmed 2000). Coloniality demonstrates both continuity and discontinuity because it occurs across space in variegated and dynamic ways. For the Caribbean decolonial thinker Nelson Maldonado-Torres, coloniality saturates 'so many ... aspects of our modern experience. In a way, as modern subjects we breathe coloniality all the time and every day' (2007: 243). For decolonial geographers, coloniality is encountered in every *location* and *place*. As modern subjects, we live and work in, move through and occupy, *spaces* that are shaped everywhere by coloniality.

Coloniality refers to enduring relations of power and difference that have, since the late fifteenth century, configured routines of exchange, encounter, rule, identity, exploration, imagination and sense of self. These routines have been codified in institutions, interactions and narratives passed from one generation to the next. In material forms, coloniality has systematically deployed different types of violence that result in racial, class and gendered oppression and psychological harm (**section 1.III**). Because of its institutionalization, codification and violence, coloniality has, it is argued, become inseparable from modernity. Coloniality comprises the dark side of modernity (Mignolo 2000), with processes in the present linking back discontinuously to processes from the past.

Yet, for many, coloniality is just the way the world works. Due to the profound restructuring of the world after the conquest of the Americas – what critics term the catastrophe, or the Columbian legacy of genocide – relations of economy, politics and society were transformed, and became interlinked in ways that favour western Europe and North

America. The power relations of coloniality rested – and still rest – upon the articulations of race, labour, space and peoples to the benefit of capitalist institutions, white populations and north Atlantic metropolitan countries (Quijano 2000; Escobar 2007). The pursuit of power and wealth entailed the displacement and separation of Black, Indigenous and diverse subordinate populations from their territories and communities. These subordinated groups' ways of living in and knowing about the world have also been radically restructured, through the imposition of Eurocentric social, religious, ideological orders and rules of social interaction. Hence coloniality operates 'all the way down', from globe-spanning economic relations through to the books we read. Coloniality thus refers to

> longstanding patterns of power that emerged as a result of colonialism but that define culture, labour, inter-subjective relations and knowledge production well beyond the strict limits of colonial administration. Thus, coloniality survives colonialism. It is maintained alive in books, in the criteria for academic performance, in cultural patterns, in common sense, in the self-image of peoples, in aspirations of self. (Maldonado-Torres 2007: 243)

In the colonial-modern world, racial difference and hierarchy are fundamental to relations of power. Western concentrations of wealth and territory have been constructed through the exploitation of enslaved and indentured labour, and the dispossession of groups designated as racially inferior to white Europeans and **settlers**. Racial hierarchies shape the spaces and the nature of social encounters, and consolidate unequal access to formal education, property and social standing. Coloniality is intricately connected to a **racial capitalism** by which market-driven production, labour and control of capital co-developed through enslavement, genocide, and resource and territorial appropriation.

One facet of coloniality's operation is the discomfort felt when colonial power is made visible. Colonial-modern worldviews deny that colonialism is present now and every day, and instead attribute it to the past or even take pride in that past (**Box 1.1**). Acknowledging that coloniality is present now and everywhere *is* unsettling, as it challenges those who benefit from coloniality to face it. Colonial power and domination are not discussed openly and thoroughly in national debates or in geography, and frequently cause the voices of those affected by racism and dispossession to be less audible. Consequently coloniality is treated complacently, relieving the privileged minority from becoming accountable. One key step in decolonizing then involves stepping into and inhabiting that discomfort and recognizing that, in comparison with coloniality's injustices and truncation of lives, a sense of discomfort is manageable. Concretely, decolonizing is furthered by taking responsibility for the *causes* of discomfort, and using that discomfort to prompt decolonial action (Ahmed 2000; de Leeuw and Hunt 2018).

Box 1.1 British opinion on colonial histories

In January 2016, the UK-based market research firm YouGov found that 29 per cent of respondents agreed that 'Britain tends to view our history of colonisation too positively', while around the same share (28%) declared Britain's colonial history was viewed too negatively (the remainder felt the balance was about right). In the same poll, three-fifths thought the statue of Cecil Rhodes at the University of Oxford should not be taken down, despite widespread student mobilization (Chantiluke et al. 2018). In 2016, young people aged eighteen to twenty-four years were the most likely to hold negative views of colonial history (40 per cent) (YouGov 2016). In a 2019 survey across selected European countries and Japan, around one-third of British respondents viewed the empire more as a source of pride than shame, second only to the Netherlands (YouGov 2020). Meanwhile in Oxford, the Rhodes statue remains in place.

As the textbox on colonial histories suggests, coloniality's 'taken for grantedness' rests in part on attitudes and the types of knowledge people hold about what modernity is. Decolonial commentators argue that coloniality imbues modern modalities of thought that become dominant through key institutions and processes. One domain is education, as students, instructors, books, and theories circulate in schools and universities and out into society. Together they reproduce the dominant ideas about how the world works (de Sousa Santos 2014; Cupples and Grosfoguel 2018). Formal education coexists with influential media, communities and public debates that shape customary understandings, assumptions and social attitudes. Throughout late modern colonialism, from the eighteenth century to the twentieth, English, French and other European languages, knowledges and criteria of academic excellence were institutionalized in colonies (**Box 1.2**). As section 1.IV shows, geography was entangled with imperial projects and colonial schooling, fashioning its current racial make-up, **curriculum** and teaching (**Chapter 5**). Because of these processes, the geography discipline has only recently begun to take responsibility for and challenge coloniality.

Calls to decolonize are heard across the world today. This chapter places decolonial action across society in dialogue with processes to decolonize academic geography. From the Dakota pipeline protests to Māori claims over rivers, decolonizing brings together diverse voices from multiple geographies, histories, ecologies and socio-spatial relations. Dissent from and resistance to colonial logics have existed since the European conquest of the Americas in the late fifteenth century (Young 2003), while action to dismantle the colonial aftermath in imaginations, social relations and territorial-ecological relations began over 500 years ago (Galeano 1971). In this respect, geography – whether in South Africa or the UK, Australia or Canada – needs to move beyond acknowledging the discipline's role in colonialism to confront its colonial *present* and address the 'active nature of geographical knowledge in sustaining colonial relations' (Holmes et al. 2014: 541). Yet decolonizing is

Box 1.2 Resisting European knowledge systems in Africa

Since the 1800s, African and African-diaspora thinkers have developed a number of critical approaches to European, colonially imposed worldviews. These include Garveyism, Ethiopianism, Negritude, Pan-Africanism and Afrocentricity, each with their distinctive criticisms of Eurocentric thinking. Diverse Black political and intellectual contributions have challenged the **coloniality of power** which exists differentially across Africa (Sihlongonyane 2015). In post-independence Tanzania, for instance, President Julius Nyerere adopted the non-tribal, non-European language of Swahili to counter colonialism. In post-independence Kenya, academics argued for the abolition of the University of Nairobi English Department (which taught selected British texts) and its replacement with African-language programmes (wa Thiong'o 1995; Barnett 2020). Scholars Ngũgĩ wa Thiong'o, Henry Owuor-Anyumba and Taban Lo Liyong campaigned for African literature and languages to be included on reading lists (wa Thiong'o 1995). The exclusion of speaking and writing in local languages, they argued, meant that unique non-western expressions and relations with the world were being lost. Wa Thiong'o advocated 'decolonizing the mind', starting in universities whose teachings naturalize and validate coloniality (wa Thiong'o 1986).

not straightforward, as recent media coverage highlights. Colonial institutions, attitudes and practices push back against decolonial action and thinking, adaptively reproducing the unquestioned dominance that benefits some, while truncating lives and degrading places elsewhere. For this reason, decolonizing generates moments of revelation and feelings of discomposure. It is not about abstract theoretical debates. It is a discomforting journey for individuals and institutions to realize that what we (here meaning English-speaking, mostly white readers in western-type universities) think and do contributes actively and materially to coloniality.

In summary, coloniality comprises concrete and enduring processes arising from the configuration of European forms of modernity that gained hegemonic power through expansive and deliberate measures to further wealth concentration, political control, and the organization of space from the fifteenth century to the twenty-first. To fully understand coloniality and decolonizing requires looking at the geographically variable patterns of colonization and the outcomes of colonial and imperial power in territories around the world.

III Historical Geographies of Colonialism and Decolonization

Having outlined the core features of coloniality, we can now ask where and how coloniality became established. Many decolonial and anti-colonial scholars agree that a major change in worldwide economic, political and socio-cultural relations began to gain momentum from the late fifteenth century. That turning point saw southern (later western) Europe write the rules of the incipient world system and, through plunder and violence, begin to accumulate wealth and geopolitical power. In turn, the economic and political system was consolidated through practices, administrations and institutions that shaped the world around European mentalities and interests. This section provides an introductory overview of this history from the fifteenth century to the twentieth. The west's rise as a globally powerful influence was integrally bound up with the creation of different types of colonialism, largely but not exclusively overseas (Quijano 2000). The section also explains why formal political independence in the former colonies did not end coloniality, which continues in both **postcolonial** countries and **settler colonial** states.

Prior to the fifteenth century, there existed world-spanning interconnected circuits of trade and manufacture centred around hubs in the eastern Mediterranean, China, India and Europe. Europe was a relatively minor, marginal sphere in these wider circuits, which worked through the Mediterranean sea and across the Silk Road, using sophisticated credit and navigation systems (Abu-Lughod 1989). This had changed markedly by 1600 as Spain and Portugal, followed by northwest European countries, embarked on creating an unprecedented political-economic-social world system (Abu-Lughod 1989; Sidaway 2000). Between 1250 and 1350, this incipient world system existed with broadly similar levels of development across regions. After 1350 however, Europe devised new rules for outward connections which were based on plunder and religious intolerance and Atlantic-oriented trade (Abu-Lughod 1989). Turning towards the Atlantic ocean, Europe began to institutionalize extractive and expansive structures of African slavery, plantations and colonial administration, leading to the systematic reorganization of power in the world (Quijano 2000). **Doctrines of discovery** based in European law legitimated annexation and colonization. Different forms of European colonialism emerged, forging European dominion in Africa, the Americas and Asia and effecting worldwide transformations.

Colonialisms: plural forms, diverse outcomes

Colonialism refers to the control of people and territory by another country, which often justifies its actions with ideologies that invoke civilization or racial superiority. Colonization and colonialism refer to multiple ways of invading territories and exerting dominion. Partly this multiplicity is due to the diverse origins and histories of colonialism (from the sixteenth century to the nineteenth, including French, Spanish, Dutch, Italian, Portuguese, Belgian and British colonization), which strongly influence the particular configurations of coloniality in the present, and how people understand coloniality (McClintock 1992; Young 2001). Another factor is the varied forms of administering colonized territories and resources and of controlling local populations.[3]

[3] Accessible introductions on colonialism can be found in Gregory et al. 2009, Castree et al. 2013 and Young 2003.

Colonization and colonialism include projects to extract wealth as well as establish territorial domination and settler colonialism. Enclave economies operated through key transport hubs for trade in enslaved populations, and plundering gold and silver, and later through export trade oriented to European markets and profits (Galeano 1971). Imperialist Portugal and Spain connected their colonies in Central and South America, and in portions of north, west and southern Africa, with South and Southeast Asia, including the Spanish Philippines, while other European imperial ventures occurred in Africa and Asia. In settler colonies, Europeans were incentivized to move permanently to colonized territories to establish western forms of agriculture, society and politics. Significant settler colonial states today include the USA, Canada, Australia, Aotearoa-New Zealand and states in southern Africa (Cavanagh and Veracini 2016; **Chapter 2.IV**). Another colonial form comprised the systematic reorganization of colonized populations and territories around the creation and extraction of resources destined for Europe.

In each case, European colonialism was realized and maintained through violence, from the genocide of native populations through to deliberate disruptions of subsistence, spirituality, family relations and prior forms of governance. Over the centuries, Europe itself experienced the domination of empires, for example the Ottoman (fourteenth to twentieth century) and Hapsburg (thirteenth to twentieth century) empires. However, these empires did not systematically adopt the forms of violent restructuring and wealth capture found in Europe's overseas colonies.[4]

Like the colonized overseas territories, these European empires were subjected to colonizers' models for organizing life, and the imposition of selective religious, cultural, intellectual and social relations. Diverse forms of colonialism in turn adopted varied types of administration, law and governance, economy and labour, territory and property, and narratives about the reasons for colonialism. Across varied forms and

---

[4] This is not to say, however, that decolonizing is irrelevant in Europe, as the Bristol example shows; see also **Box 1.3**.

periods of colonialism, economic activity, political governance and socio-cultural relations were fundamentally restructured. The extraction of resources and wealth occurred through the occupation of lands for mining and agriculture, with profits flowing to settlers and metropolitan capitals, as colonists monopolized economic exchange, labour relations and trade. Enslaved African labour worked in plantations, cultivating crops consumed in Europe and generating profound reorganizations of ecosystems around the world. Mining was based on the coerced labour of **Indigenous peoples** in the Americas and Black Africans in southern Africa. Indentured labour systems moved Asian and other populations between colonies. Maintaining control over territories and labourers engaged the colonial powers in instituting and enforcing political and legal measures through military or other coercive means.

European colonialism sometimes devolved control over locals to existing authorities ('indirect rule'), generating novel dynamics of power and identity. Central to many forms of colonialism were forced changes to the daily life of colonized populations. For this reason, long-term colonial presence was accompanied by forced religious conversion, disruption of existing education, and the inculcation of European values and ideologies in local populations (including the desirability of European versions of modernity and development) (McEwan 2018). In this way, colonial relations of power resulted in the remaking of spatial relations between peoples, lands and environments. Together, these processes generated the complex geographies of colonialism that are visible and experienced today.

Decolonization: political independence?

Decolonization comprises the struggles of populations and their leaders to gain political independence as sovereign states from colonial masters. Decolonization has been an anti-colonial goal since late fifteenth century, expressed in 'anti-colonial dissent' manifested in words and actions (Blunt and Wills 2000: 167). Political independence was achieved in Haiti, for instance, through a slave rebellion led by Toussaint L'Ouverture in 1791, which defeated French, Spanish and British colonial troops. Decolonization entails the material and ideological contestation of colonialism, expressed in institutional and infrastructural objectives as well as anti-colonial thinking and visions for non-colonized futures. However, as discussed below, formal independence does not necessarily materialize these liberatory visions. Haiti was forced for over a century to pay reparations to European colonial powers for plantation and slavery revenues.

After Haiti, decolonization occurred at several points, with major transitions in Latin America in the early nineteenth century, and a 'second wave' of formal decolonization in Asian and African countries during the 1950s and 1960s. The term decolonization dates from the 1930s, and gained increased visibility as territories in Southeast Asia, North Africa, sub-Saharan Africa and the Caribbean achieved independent status. In practice, however, the official end of colonization gave rise to different formations of domination, racial exclusion and territorial dispossession that remain significant in these postcolonial countries. 'Postcolonial' here does not indicate the *end* of colonialism, but the condition of living with the consequences and some features of colonialism (see below). In some cases, imperialism followed hard on the heels of colonialism. Philippine and Cuban independence from Spain in the late nineteenth century was followed by US geopolitical influence in these countries (Slater 2004). In South Africa, nominal independence led to the violent anti-Black rule of white settlers under apartheid. In North American and Australasian territories, independence meant little for First Nations and Indigenous groups whose status did not change (Young 2003). In the interwar period, colonized territories were passed between European empires, thereby postponing **self-determination** in de facto colonies such as Namibia, Iraq, Syria and Palestine.

Overall, the paths to formal or nominal independence were variable in their duration, depth and nature. Mid-twentieth century decolonization frequently involved civil war, violence, ethnic cleansing, negotiations and diplomacy – and, occasionally, orderly exits (Young 2001):

In some African colonies, colonization was barely accomplished, and resistance movements of varying degrees of organization and institutionalization attended the entire colonial project. In other cases, an organized anti-colonial and nationalist movement came late, accompanied by a rapidly-assembled set of political negotiations [where] the metropolitan power wished to hand over the reins of power with utmost expedience. In others, it took a war of liberation, a bloody armed struggle by leftist guerrillas or nationalist agitators pitted against white settlers or intransigent colonial states. (Watts 2009: 145–6)

Decolonization led to the planned or de facto reorganizing of societies and geographies, and yet it remained profoundly unequal. It may have changed the rulers and the flags, but in practice (neo-)colonial rules and hierarchies continued, while economic and political extraction endured. What quickly became apparent was that colonial power and its legacies did not fade after independence, since extractive, violent and ordering patterns had become ingrained in economic relations, socio-racial hierarchies, modes of thinking and the priorities of postcolonial and settler colonial states (Loomba 2005). Colonial-modern rule bequeathed a series of structural and ideological relations that countries have found hard to jettison. Former colonies – and the territories of unrepresented peoples and Indigenous nations – continue to operate under lopsided international relations and global economic systems (Slater 2004). Economic specialization locked in their reliance on global export markets with disadvantageous terms of trade, cemented by unequal relations of labour, migration, aid and investment. Culturally too, metropolitan-oriented relations and identities have shaped religious structures, aesthetics, education and language in these former colonies and territories (Jazeel 2019). Colonialism's political institutions and ideological structures have proven remarkably resilient, and continue to be characterized by coercion, violence and force.

After colonialism comes ... almost the same but not quite, due to 'continuities from the colonial period as well as breaks from it' (Sharp 2009: 5). Nation-states embarked on political, economic and social transformations in contexts that blur neat separations between the colonial era and the era after colonialism. 'From this experience characterized by hybrid forms of identity, blurred boundaries and contradictory practices, the process of decolonization must necessarily look more complex than simply self-rule managed from above by the colonial state or mobilized from below by nationalist forces' (Watts 2009: 147). In postcolonial and settler colonial countries, legal systems, religions, languages, urban layouts and economic relations were still often Eurocentric in intent, form and purpose. Variable after-colonial politics, economics, culture and social relations are postcolonial in the sense that the (colonial) present cannot be understood except in relation to colonialism. The historian Ann Stoler describes the **postcolonial condition** in the following terms:

> the political life of imperial debris, the longevity of struc-tures of dominance, and the uneven pace with which people can extricate themselves from the colonial order of things [by means of] creative, critical and sometimes costly measures ... [The postcolonial condition entails] the production of new exposures and enduring damage. ... Modernity and capitalism can account for the left aside, but not *where* people are left, *what* they are left with, and what *means* they have to deal with what remains. ... Imperial ruins ... are less sites of love than implacable resentment, disregard, and abandonment. (Stoler 2008: 193, original emphasis)

Added to this, geographers highlight the *spatial* dimensions of the postcolonial condition, the differentiated human-geophysical landscapes with their varied socio-political, ideological, material and environmental imprints.

Decolonizing: unfinished business

Material and epistemic struggles against coloniality took place against the backdrop of the Cold War, with former colonies attempting to break from US or Soviet **hegemony**. 'Third World' countries devised an agenda for action distinct to that of the capitalist and communist blocs, forming a Tricontinental political alliance at the 1955 Bandung conference (Young 2001). At Bandung, the Caribbean writer Aimé Césaire articulated common anti-colonial grievances:

> My turn to state an equation: colonization equals 'thingification'. I hear the storm. They [the colonial powers] talk to me about progress, about 'achievements', diseases cured, improved standards of living. *I* am talking about societies drained of their essence, cultures trampled underfoot, institutions undermined, lands confiscated, religions smashed, magnificent artistic creations destroyed, extraordinary *possibilities* wiped out. (Césaire 1972, original emphasis)

Inspiring his pupil Frantz Fanon, among others, Césaire established that colonial talk of civilization was hypocritical, and that while colonialism brutalized both colonizer and colonized, the latter bore the brunt of **dehumanization**. Césaire also shone a critical light on the *interrelations* between economy, politics, society, knowledge and imagination in colonialism's aftermath – a key point for decolonial debates.

Decolonization is 'unfinished business'. And as unfinished business, decoloniz*ing* involves breaking away from coloniality's grip, a longer-term and more complex process than political independence. In light of Césaire's blistering critique of coloniality, decolonizing has involved ongoing and spatially variable actions and situations. For populations whose status after independence left them without full political standing, the challenge is to achieve recognition and rights. The Unrepresented Nations and Peoples Organization provides an international platform for nonviolent and democratic campaigns to promote self-determination. In some regions, such as in Latin America, internal colonialism persists in enclaves of racialized communities whose labour and products are cheapened through systematic domination. Another group demanding liberation and **autonomy** are the world's 476 million Indigenous peoples. While their international rights have been established under the United Nations Declaration on the Rights of Indigenous Peoples (2007), in practice these rights are unevenly implemented. Indigenous peoples in settler colonial states face ongoing exclusions and political constraints. Some settler colonial states, including Canada, Aotearoa-New Zealand and South Africa, have offered apologies for colonialism or sought reconciliation, yet pro-Indigenous structural change and peaceful coexistence face numerous hurdles. In Europe too, dominated and forcibly appropriated groups struggle for recognition (**Box 1.3**). The unfinished business of decolonizing is a profoundly

Box 1.3 Decolonizing Europe

Europe is rarely the focus of decolonizing debates, being wrongly represented as if it were a homogeneous power-house exerting dominance elsewhere in the world. In fact Europe displays a particular configuration of coloniality-modernity (de Sousa Santos 2017). Before 1492, European genocides provided templates for colonial-modern violence in the Americas (Grosfoguel 2013). Europe's colonial present has touched down harshly in Ireland and on the Sami, African diasporas and Russian peasants, among others. European Roma, who today number 11 million, were first documented in fourteenth-century slave records; since then they have been expelled, discriminated against and excluded from citizenship, giving rise to a movement for self-determination (Yuval-Davis et al. 2017). Europe also has complex colonial-modern relations with its Mediterranean neighbours, from the Catholic expulsion of Moorish rulers from Iberia, to the treatment of twenty-first-century seaborne refugees. North African and eastern Mediterranean migrant workers labour in European industries and services, yet experience systematic racism and constrained rights, which generate complex ties to homelands (Stoler 2016; Harris 2020).

political issue, as it foregrounds issues of citizenship, territory, autonomy and justice.

In summary, colonialism and its aftermaths created enduring economic, political, social and knowledge-related inequalities and forms of domination. As such, a distinction must be drawn between formal independence for colonies (termed political decoloniz*ation*) and broader decoloniz*ing* struggles. Coloniality arises out of the legacies of modern forms of colonialism, which include the incomplete and ambivalent outcomes of formal independence, and ongoing struggles for dignity, wealth redistribution, territory and self-determination. Concrete elements of coloniality's structure and dynamic can hence be identified, while colonialism's legacies hint at coloniality's capacity to rebuild its reach and power in changing circumstances.

IV Coloniality and Modernity in the One-World World

Colonial legacies saturate material (institutions, resource distribution), social (relations, attitudes, 'common sense') and mental (knowledges, presumptions) dispositions and processes. Coloniality's ongoing control results in systematic duress and inequality. Two of the factors behind this will be discussed here, reflecting current decolonial thinking. The first relates to the way unequal, hierarchical and exclusionary relations come to be regarded as 'just the way the world works'. The second factor points to the creation of the colonial-modern world in relation to European arbitration over optimal outcomes for resources, societies and institutions. These factors, although expressed differently across time and space, can be thought of as now global but originally Euro-American designs (Mignolo 2000). Coloniality is inextricably interlinked with modernity, reflected in the interchangeable terms 'modernity-coloniality', 'modern-colonial' and 'coloniality-modernity'. Colonialism created a world in the image of modern colonizers, and persuaded (often with force) the rest of the world that modernity is foundationally established. The geography discipline was a cornerstone in this construction. For critical scholars, coloniality, modernity and world capitalism co-evolved in economic-politico-social relations after 1492 through a new world system (Escobar 2007: 184; **Chapter 2.III**). **Modernity-coloniality** bridged from Iberia (and later from northwestern Europe) to west Africa, to the Americas and the Caribbean, to Asia and the Pacific. Modern-colonial patterns of domination persist today in north Atlantic political economies, white-dominated racial hierarchies and Eurocentric ideologies.

One dimension of coloniality's extension and persistence relates to how colonial-modern relations have been presented as 'just the way the world works'. The world is modern, so how can coloniality be? Decolonial commentators suggest that European colonialism reorganized relations of labour, governance, society and economy so deeply and extensively that the world now reflects that reality. Due to its dominance, coloniality has become routinized and predictable, perceived as a system which has seemingly made modernity. This has occurred through territorial, political and socio-cultural organization and routines that reflect the values and priorities of elites (Wainwright 2010; **Chapter 2.III**). Although coloniality produces sharp injustices and undermines lives and environments, it is upheld by powerful institutions, narratives and distributions of property and social capital. Colonialisms operated through force and violence, yet modernity has been consolidated through the seemingly self-evident superiority of Eurocentric systems and operations. Coloniality becomes hegemonic as it establishes the taken-for-granted nature of modernity.

The second factor in coloniality's interlocking with modernity relates to the Eurocentric criteria that arbitrate over outcomes for resources, societies and institutions. Eurocentric systems for knowing and organizing the world have been oriented to western notions of optimal effectivity and outcome. In early modern Europe, developments in philosophy, exploration and experimentation reinforced coloniality's enduring power in shaping material and ideological relations. In the fifteenth century, mechanistic modes of science marginalized other European forms of enquiry and became 'particularly unresponsive to non-western ways of thinking about and interacting' with the world's diverse societies (Adas 2016: 605). Crucially too, western science was subordinated primarily to colonial imperatives of value extraction and security. Europe's metropolitan science adjudicated on what counted as worth knowing or doing, and came to believe its findings had worldwide applicability. Explorers and scientists tended to treat 'local', non-western practices and knowledges as limited or as obstacles to colonial ambitions, even as they appropriated some of those knowledges (Tuhiwai Smith 2006).

These practices and presumptions bolster what the historian of science John Law terms a '**one-world world**' (Law 2015) that has sealed itself against other knowledges and realities. Exploration, early science and renaissance classification systems together informed a new understanding of the world as if it were a single entity that was run optimally and effectively with universally valid systems of knowing and organizing. Science sought the underlying 'laws' of nature, but what we now call science developed in relation to a particular set of actors and institutions during a major upheaval in political, economic, social and intellectual relations. In this domain, the findings of explorers, thinkers and amateurs were organized and validated through Eurocentric practices and institutional arrangements, and codified within their logics of interpretation.[5] Unsurprisingly, the European sciences began to believe their perspective was 'above' the world, having gained universal reach and applicability. The seventeenth-century French philosopher René Descartes understood this universal reach as an ideal modern viewpoint, a 'God's-eye view'; in the eighteenth century, the German philosopher Immanuel Kant argued that 'categories of space and time [are] innate to the minds of [white, western]

[5] During colonialism and beyond, practitioners of 'Other' forms of knowing helped western scientists collect samples and information. These dynamics are examined in the Royal Geographical Society's exhibition 'Hidden Histories of Exploration'. See https://www.rgs.org/geography/news/hidden-histories-of-exploration.

"men" [*sic*]' (Grosfoguel 2017: 149). Emerging in opposition to racial and cultural Others, European self-identity rested upon the celebration of an unprecedented secular, scientific and rapaciously expansive system (Wynter 2003; Lugones 2007).

Europe crafted the one-world world doctrine and exported it to its colonies (Quijano 2007). Backed by force and administration, western science came to define modernity. Knowledges *about* the world were systematically collated to inform colonial and imperial *interventions* in economy, politics, settlement, colonization and society across the world. However, European systems of organizing and knowing the world did not go unchallenged (**Box 1.2**). In the 1950s, Césaire rejected the colonial hierarchical binary between a supposed **universalism** and the supposedly narrow non-European ways of organizing and knowing the world (Grosfoguel 2017). Decolonial commentators have furthered this critique, showing how European designs were extended to the detriment of other life-worlds (Mignolo 2000; de Sousa Santos 2017). The concept of coloniality thus contributes 'to a better understanding of the power and knowledge structures of the system we have inhabited for the past 520 years' (Grosfoguel 2017: 162).

The one-world logic brought with it categories and classifications that reflected European goals and assumptions. These categories arise from and validate one perspective, being imbued with powerful blinkers and biases. For instance, feminist geographers have shown how language conveys value-laden gendered associations of suburban and rural landscapes (McDowell and Sharp 1999). Decolonizing discussions alert us to how colonial-modern categories and meanings attributed to places and peoples become dominant, despite their selectivity and anti-colonial resistance to them. Take 'Earth' – a seemingly straightforward term for a rock in space inhabited by animals, humans, plants and microbes. To a white, northern reader this is a straightforward definition. Yet it is neither valid nor relevant universally. Australian Aboriginal groups conceive of the Earth as a life-world in which humans are integrally and fundamentally brought into being through the same processes that create geology, nature and water in a continuous development (Bawaka Country et al. 2016a). In the Andes, guinea-pigs, mountains and humans are not separated into discrete categories (animal, mineral, human) but are dimensions of the *ayllu*, interconnected beings with reciprocal relations for survival (de la Cadena 2010). According to Eurocentric universal criteria, Aboriginal and Andean systems represent cultural or religious beliefs, not organized and internally coherent *knowledge systems*. Under coloniality, Andean and Aboriginal – or Roma, Black … – knowledges cannot represent a world unto themselves. This marginalizes non-western knowledge-holders, including Aboriginal, Indigenous and other scientists (**Chapter 3; Chapter 4.IV**).

Physical geography and colonial science

As it is based on modern scientific practices and institutions, physical geography is linked broadly to this history of colonial science. Physical geographers have recently begun to reflect on the implications of colonial-type science for people, modes of enquiry and the social institutions that facilitate their work (Lave 2015). Physical geographies are constantly developing new findings, experiments and valuable outputs. Decolonizing asks questions about why certain practices and approaches are taken for granted, and how physical geography might engage with a more diverse set of knowledges to understand its objects of study. Acknowledging the colonial mode of science and its extension across the world does not entail discarding physical geographical science or viewing it as irrelevant. Rather, decolonizing argues that recognition of the impacts of colonialism and coloniality, and openness towards non-western or non-'modern' practices, knowledge-holders and forms of experimentation, can go a long way towards recalibrating historic hierarchies and exclusions (see **section 1.VI**). As this book argues, physical geography stands on the threshold of a decolonizing mode of enquiry which seeks to undo colonial legacies.

Geography's 'one-world world'

Geographical skills and institutions contributed integrally to colonial systems of knowledge and power, thereby helping to build a world around European colonialism. The geographers' toolkit concretely facilitated extraction, territorial dispossession and the restructuring of non-European lifeworlds. Mapping, surveying, exploration, navigation, town planning, resource identification, plantation agriculture and other geographical tasks enacted colonial-modernity across the globe. Although in the early colonial-modern era geographers were not professionals, by the nineteenth century geography was tightly and institutionally connected to imperialism (Driver 2001). Presenting itself as an applied discipline, its status rested upon its claim to know the world through exploration and fieldwork. Colonial authorities used sciences of geography and botany as field-oriented 'frontlines' in political-economic projects of wealth extraction, while ethnography assisted colonial surveillance and control (Adas 2016; Tuhiwai Smith 2006). Geography applied its knowledge to make colonialism more efficient, as for example with Greenwich Mean Time, measured from a south London military base from 1675, and the international time zones introduced in the late 1800s (Gregory et al. 2009).

Geography also contributed to dominant explanations and justifications for the expansion of colonialism. From the fifteenth century, a 'governing vision' presumed that European minds and rationality created progress and innovation, which flowed to non-European areas and cycled material wealth back to Europe. The geographer J.M. Blaut argued that this 'diffusionism' originated in 1492. At this moment geographical divides between Iberia (later all of Europe) and the colonies encouraged western powers to think that exploitative expansion was inevitable and justifiable on religious (and later secular) grounds (Blaut 1993; Wynter 2003; Slater 2004). Over the centuries, diffusionism was adapted to global situations, attributing superiority of mind, civilization and 'scientific' racial factors to colonizers (Livingstone 1992). People using geographical methods accepted diffusionism's values, convinced that metropolitan law, cartography, property and land use were universally constructive. With the help of geographers, colonizers emptied spaces of their indigenous human, animal and vegetable inhabitants and filled them with European plants, animals, settlers and languages (Howitt 2001a). In the first half of the twentieth century, geographers theorized diffusion in terms of spatial laws, ignoring the (neo-)colonial consequences for societies and landscapes (Gregory et al. 2009: 160–2). In Blaut's (1993) memorable phrase, diffusionism was 'the colonizer's model of the world'.

Diffusionism was accompanied by negative comparisons of non-European societies and places in relation to Europe, a spatial ranking that endures to the present (as in 'developing vs developed countries'). Colonial-modern Europe provides the 'standard' against which Others were and are judged wanting, which in turn justifies interventions.

'The short hand is that patriarchal, colonial, and imperial legacies continue to inform the discipline of geography; the theoretical and methodological purpose of the discipline is twinned with exploration and conquest and European masculinist ways of knowing' (McKittrick 2019: 244). **Eurocentrism**, in these terms, rests upon systematic institutional power, as a 'matter of science, and scholarship, and informed and expert opinion' (Blaut 1993: 9). For example, British colonial organization of space in India reverberates today through urban planning expertise, university geography teaching and an urban aesthetics which makes western-style cities more desirable (Sundaresan 2020).

As the underside of modernity, coloniality is a modality of power, knowledge and being that operates in the present. Coloniality underpins white supremacy and Euro-American power, denying validity and voice to experiences and thought outside its realm. Nevertheless, a one-world world and Eurocentric comparisons are neither inevitable nor homogeneous, due to resistance and institutional change (**section 1.VI**). Before turning to discuss decolonizing however, we will address the corrosive effect of racism and racial hierarchies in the world – and in the geography discipline.

V Racism in the Colonial Present

As mentioned above, coloniality as a matrix of power rests upon racial hierarchies that are inherent in modern capitalism, while non-European peoples and knowledge systems have been systematically subjected to violence and assimilation. Race is a social construction which reflects social (colonial-modern) categories and distinctions that exist in historical and geographical contexts. The events that culminated in the upending of Colston's statue into Bristol harbour reflect over 500 years of colonial-modern racial hierarchies and continuous collective action to overturn structural racism. Coloniality is deeply racialized, meaning that race is always present in modern institutions and structures. Racial disparities bring material benefits to those identified as white, whereas non-white groups carry multifaceted costs.

> ### Racialization
>
> Racialization denotes processes that institutionalize and maintain disparities between socially determined categories of difference ('race'). Racialization is systematic, encompasses social, political, economic, psychological and epistemic relations, and informs racism.

From the eighteenth century onwards, the notion of 'race' and racial categories co-emerged with coloniality and modernity. Modern-colonial racial categories arose to define the groups to be conquered, enslaved and put to work during American colonization and beyond. During the Catholic re-conquest of Spain, political status distinctions between Moors, Christians and Jews based on notions of 'blood' had already been devised. Later colonial racial classifications informed Latin American divisions of labour that allocated Indigenous peoples to mining, weaving and domestic service, and enslaved Black Africans to plantations and urban services (Quijano 2000).

Coloniality continues to deploy **racialization** to organize economic, political and social relations, and to justify violence against non-white, non-European bodies and places (Kobayashi and de Leeuw 2010; Maldonado-Torres 2016). Racial capitalism continues to differentially reward working bodies and communities, while environmental racism is evident in non-white groups' disproportionate exposure to pollution and harm (Pulido 2017). Racialization exposes non-white groups to disproportionate risk, dehumanization and removal of care, resulting in material and life-determining effects in segregated neighbourhoods, malls and universities (Inwood and Yarborough 2010; Roy 2016). Racialization is inherently spatial, with domination operating across scales. Although **whiteness** operates globally across diverse spaces, the nuances of racial difference vary in their expressions and structures. For example, being Black is experienced differently not only in Harare and Chicago but also *within* those cities, while **anti-Indigenous racism** is expressed differently to anti-Black or anti-Muslim racism. Racialization is a dynamic socio-spatial process, constructed in 'the relationship *between places*' (Pulido 2000: 13, original emphasis). Racialized spatial arrangements and social interactions determine when, where and how non-white bodies are considered to be 'out of place' in a world constructed around whiteness (Fanon 2008; Ahmed 2007).

Racialization thus focuses attention on the – often unspoken – power of whiteness. Whiteness and its associated privileges are invisible to groups identified as white (primarily Europeans, their descendants and colonial settlers) (Ahmed 2007; Bonnett 2014). Under colonial-modern racialization, whiteness is the implicit norm, what Reni Eddo-Lodge (2018) terms an 'invisible monolith'. Despite the unremarked power of whiteness, white experiences, knowledges, values, institutions and intersubjective relations are naturalized (Esson et al. 2017: 386). The very appearance of whiteness as neutral confers supremacy (Bonnett 1997; UCL Collective 2015) and grants facilitating advantages:

> I have come to see white privilege as an invisible package of unearned assets that I can count on cashing in each

day, but about which I was 'meant' to remain oblivious. White privilege is like an invisible weightless knapsack of special provisions, maps, passports, codebooks, visas, clothes, tools and blank checks. (McIntosh 1989: 1)

As Peggy McIntosh notes, whiteness bestows *spatial* privileges. With geography so close to coloniality, the notion of race was central to the discipline from the nineteenth century to the early twentieth. Imperial geography mapped the spatial distribution of 'races', obsessed about acclimatizing white settlers in the tropics, and made moral judgements about civilization and degeneration (Livingstone 1992; Kobayashi 2002). Although colonial geographies of race faded in significance through the twentieth century, the **unmarked power** of whiteness to determine the scope and focus of the discipline has persisted. As critical geographers have documented for over twenty years, racialization results in unacknowledged norms of whiteness and the exclusion of non-white groups (Peake and Kobayashi 2002; Bonnett 2014; Mahtani 2014; **Table 1.1**). In this context, the Chicana geographer Laura Pulido argues that whiteness 'skews our intellectual production' (Pulido 2002: 45). *All* geographers – Black and white, physical and human – are bound up with the distorting power of whiteness. Although largely unconscious, this produces **'white geographies'**, that is 'ways of seeing, understanding and interrogating the world that are based on racialized and colonial assumptions that are unremarked, normalized and perpetuated' (Domosh 2015: 1; see also Dwyer and Jones 2000). White ways of seeing the world implicitly categorize certain places and populations as racialized. Seemingly neutral terms such as 'countryside', 'wildness' or 'nature' carry colonial connotations and imaginative geographies. For instance, countryside is associated with whiteness, whereas wildness is associated with disqualifying characteristics of non-white (Indigenous, Black) groups (**Box 2.3; Chapter 4.IV**). Recognizing and changing such associations is an ongoing challenge. The *National Geographic* magazine recently acknowledged its racializing representations of places and peoples (Goldberg 2018).

Table 1.1 Racial disparities in UK and US geography
The table provides a provisional snapshot of geography's racialization.
The categories shown reflect official classifications. The UK 2011 census
records 14.1% 'Black, Asian and Minority Ethnic' (**BAME** comprising
7.5% Asian, 3.3% Black and 3.2% mixed and other) and white 86%.
The USA records 60.1% white, 18.5% Latinx, 13.4% Black/African
American, 10.2% Asian/Indigenous/Pacific/mixed race. Information
sources relate to 2012, 2015 and 2005–15 data.

	UK geography	UK all disciplines	USA geography	USA social sciences
STUDENTS				
PhDs by non-white (US)/BAME (UK) students	4.4%	16.4%	8%	19.5%
BAME undergraduate students	6.3%	>20%		
FACULTY				
White, non-Hispanic			71% overall, 91% of full professors	
'International'			15%	
BAME	4.3%	8.2%		
PROFESSIONAL ASSOCIATION				
Hispanic			4.3%	
African-American			3.15%	

Source: Adapted from Desai 2017; Faria et al. 2019; Ybarra 2019; and
Derickson 2017.

Systematic racism in the discipline, Black and anti-racist
geographers argue, dismisses and devalues the experiences,
histories, spaces, knowledges and resistance of racialized
groups (Johnson 2018). This negatively impacts the minority
of geographers who are not white, especially women of colour
(Mahtani 2014; Tolia-Kelley 2017). Despite critical anti-racist
voices, geography remains a structurally racist and norma-
tively white discipline. For this reason, actions to overhaul
geographical institutions and thinking are absolutely central
to decolonizing (Kobayashi and Peake 2000). Anti-racism is
not the responsibility of geographers of colour, as the discipline is largely white and hence accountable for removing
racism (Daigle and Sundberg 2017). In the USA and Britain,

university geography departments routinely recruit disproportionately more white than non-white students and faculty,
in respect of national averages and in comparison with other
disciplines (Pulido 2002; Desai 2017). In the USA, 'despite
decades of recognition and worry ... [geography] remains
persistently white ... dominated by white bodies and ...
norms, practices and ideologies of whiteness' (Faria et al.
2019: 364). Similarly, British geography 'fails to attract a
diverse BAME student population, which then fails to support
a pipeline of BAME staff for the discipline' (Desai 2017: 322).

   Despite and in face of these challenges, Black, Indigenous
and anti-racist geographers document and critically analyse
the spatial logics of racialization and colonial-modern
racism 'from within' and thereby document **geographies
of racialization**. Their work, emerging against geography's
overwhelming whiteness, profoundly unsettles white theories
and structures (McKittrick and Woods 2007; Johnson
et al. 2007; **Chapter 3.IV**). In this way, Black, Indigenous,
anti-racist and other geographies of colour provide highly
significant and necessary insights for decolonizing geography.
Yet, as the core problem remains unacknowledged white
norms, anti-racist practice and delinking the discipline as a
whole from structural racism is primarily the responsibility
of white geographers. With this in mind, the next section
outlines the features and directions of decolonizing physical
and human geographies.

VI Decolonizing Geography: An Introduction

As the above discussion makes clear, decolonizing denotes
a direction of travel beyond political independence, toward
the critique of and action against racialized violence and
Euro-American claims over resources and knowledge. For
the Māori scholar-activist Linda Tuhiwai Smith, decolonizing
comprises 'a long-term process involving the bureaucratic,
cultural, linguistic and psychological divesting of colonial
power' (Tuhiwai Smith 2012: 33). Divesting from colonial-
modern power involves action and thinking to undo and

delink from the **colonial matrix of power**. By disobeying the strictures of coloniality, decolonizing opens up avenues to imagine, enact and construct alternative modernities. Decolonizing – also termed **decoloniality** – encompasses material and epistemic dimensions, transforming institutions, structures and social relations, and modes of thinking. Geography has a long way to go to fully acknowledge and address coloniality, which remains an 'inconvenient truth', as Tariq Jazeel (2017) notes. Decolonizing, however, is a hopeful process, being 'always adaptive and unfixed' (de Leeuw 2017b: 311; **Chapter 3.II**).

For this reason, decolonizing is an anti-colonial praxis for living differently in the world, while thinking innovatively about how to bring about change (Mignolo and Walsh 2018; McKittrick 2019; A. Davies 2019). Due to the open-ended and unprecedented basis for decolonizing, existing frameworks and politics need to be 'stretched' to facilitate new actions and new analyses (Fanon 2004; Scott Lewis 2018). Geography's moves towards decolonizing can be traced back to the 1960s, when scholars nuanced the discipline in relation to colonialism, imperial institutionalization and resistance dynamics within and across colonized and metropolitan spaces (Driver 2001; Sharp 2009; **Chapter 2**). David Slater (1992) advocated 'unlearning' Euro-Americanism and learning from Latin American and African perspectives to understand international geopolitics (Slater 2004).

Decolonial praxis is also associated inextricably with ethical commitments and responsibilities that seek to bend institutions towards inclusivity and create spaces for respectful treatment, dialogue and just distributions between life-worlds. Decoloniality represents more than a **paradigm shift** (such as geography's cultural turn in the 1990s) as it encompasses political, institutional and personal actions at numerous scales (Mignolo and Walsh 2018; Maldonado-Torres 2011). Responding to decolonizing goals requires flexibility and a willingness to engage with critics (Bhambra et al. 2018; Pickerill 2018; **Chapter 3.V**), as existing institutions and powerful groups subtly undermine decolonial anti-racist actions (Ahmed 2012; Eddo-Lodge 2018). Decolonizing does not discard all existing geographical knowledge, methods and approaches, but reassesses them critically. It seeks decentred and co-learnt ways of thinking about, and acting to nurture, plural worlds. Decoloniality does not seek to bring the world's heterogeneity into one comparative frame, but to historicize and remake explanations from numerous locations. In place of diffusionism's single time-line from (north Atlantic) past to (one-world) present, decolonizing recognizes overlapping and interconnected geographies. For example, in heavily colonized areas of Western Sydney, Australia, there is now dialogue between Aboriginal groups and non-Aboriginal settlers to co-learn about bushfires, plant communities and conservation. Aboriginal groups bring their knowledge about healing Country (which encompasses land, birds, people, wind and spirits bound together in a dynamic whole) to realize decolonizing (Darug Ngurra et al. 2019; **Chapter 4.II**).

Decolonizing means leaning into the geography discipline's discomfort about discussing **white privilege** and racism. No longer is it possible to say whiteness is a 'neutral' positionality, given the urgent need to 'expose the white background of academia' (Johnson 2020a: 91). In light of the racial disparities in professional geography, white geographers will need to engage fully with anti-racism and decolonizing (Daigle 2019). 'The discipline of geography will retain its Eurocentricity, coloniality and whiteness unless all geographers begin to do the anti-racist and decolonial work historically done by Indigenous people, people of colour, women and queer faculty and students' (Daigle and Sundberg 2017: 340). Anti-racism is informed by interdisciplinary **critical race theory** and anti-racist practices, as well as social justice agendas, which have not received widespread attention (Roy 2016; Esson 2018). Identifying context-specific whiteness and racialization also means listening carefully to non-white faculty and students' experiences. In tertiary education, practical measures include outreach for under-represented candidates, travel bursaries, curriculum change, employment of non-white staff, and training in anti-racism and being a white ally (**Chapter 5**).

Precisely because, under coloniality, different groups speak past each other, it is the responsibility of all geographers to work and speak with plural groups outside the parameters of institutional practice (Ahmed 2000; Howitt 2001b). In this sense, the goal is one of *un*learning in order to *re*-learn through dialogues and shared practice with heterogeneous sectors. Despite the claims of universal geographical knowledge, ways of knowing and thinking about the world remain highly diverse, as well as contested and spatially variable, because the societies (and natures) informing them are similarly variegated (Haesbaert 2021). Disentangling from coloniality is helped by listening, watching and focusing on these other modalities, accompanying and 'walking with' them. Transforming praxis and how one thinks about the world – alongside a differently positioned group – unsettles preconceptions and suggests alternatives (Sundberg 2014; Mignolo and Walsh 2018; **Chapter 3**). Decolonizing is inspired by

> anti-colonial writers, including W.E.B. du Bois, Aimé Césaire, Frantz Fanon, Gloria Anzaldúa, L.T. Smith, as well as movements such as the '500 years of resistance', the World Social Forum and the Zapatistas. Decolonial approaches emerge from, and engage with, a wide range of critical and radical scholarship, including critical Black scholarship, Indigenous theory, feminist and queer theory (especially that informed by non-metropolitan concerns), and the modernity-coloniality-decoloniality (MCD) school. (Radcliffe 2017a: 329)

Within the United States, decolonizing thrives in the plurality of Chican@, Indigenous, Latinx, Pacific Islander, Black, Asian and critical diasporic perspectives (e.g. Masuda et al. 2020). For example, the Standing Rock syllabus provides students and instructors with a structured introduction to Indigenous voices and decolonial frameworks to facilitate understanding of disputes over oil pipelines and territory.[6]

---

[6] See https://nycstandswithstandingrock.wordpress.com/standingrocksyllabus.

According to the Cameroonian political theorist Achille Mbembe, the decolonial challenge is to 'rethink Africa, or for that matter, to write the world from Africa or to write Africa into contemporary social theory' (Mbembe 2010: 656; see also Comaroff and Comaroff 2012).

Decolonizing highlights the role of universities in colonial-modern structures. Campus movements such as the Rhodes Must Fall campaign at the University of Cape Town, South Africa – and later at London's School of Oriental and African Studies and at Oxford University – defy white and colonial disciplinary frameworks and exclusions (Said 1978; Chantiluke et al. 2018; Bhambra et al. 2018). Following protests, South African public universities remade their geography curricula to ensure that Black and African authors and locally relevant material were included. However, the line between colonial and non-colonial content is complicated, as Africa has plural modernities and aspires to learn from south–south exchanges (Mungwini 2013; Ndlovu-Gatsheni 2013; Knight 2018). Decolonizing raises these questions and finds that answers vary across contexts depending on existing practices, critiques and actions. Historical geography, for instance, looks very different if we start from Nigeria (Craggs and Neate 2020). Over 100 practical measures to tackle barriers to Black, Indigenous and other groups' inclusion at university were identified by Indigenous advisors in Canada (Pete undated; Whyte 2018).

Decolonizing geography is not about dismissing globalized institutions and knowledge in order to replace them with Andean, Black or African ways of life and knowledges. Rather, it involves exchanges and dialogues between plural knowledges and enacting new patterns of coexistence (Howitt 2020). Decolonizing has to be enacted and lived by building plural other worlds, located in colonial-modern geo-histories of science and enquiry. 'Decolonizing knowledge necessitates shifting the geography of reason, which means opening reason beyond Eurocentric and provincial horizons as well as producing knowledge beyond disciplinary boundaries' (Maldonado-Torres 2011: 10).

Decolonizing physical geography

Geography relies on western scientific knowledges and status in Eurocentric modes of enquiry and interpretation. Decolonizing geophysical science involves actively challenging colonialism, racism and the oppressive structures that underpin physical geography projects, themes and institutions (McAlvay et al. 2021). First, this means admitting that the 'repercussions of colonialism are as fundamental as ... plate tectonics' (Kershaw et al. 2014: 459). By implication, geophysical elements have to be understood alongside and as entangled with socio-biophysical landscape-making processes and histories of colonialism. Geographers decolonizing the geophysical processes in wetlands of Chicago for instance, acknowledge histories of colonization, Eurocentric legacies of wetland modelling, and Indigenous knowledge in order to broaden insights and logics of interpretation. In this way, physical geography can learn from non-western knowledge-holders about geophysical features and processes including water flows, soil structure and glaciers (Carey et al. 2016; Miller 2019; Dhillon 2020). Biogeography has been at the forefront in engaging with decoloniality. For example, botanical scientists at Kew Gardens, London have worked with Brazilian Indigenous communities and scientists to understand plant characteristics and uses.[7] As most physical geographers are not Black, Indigenous or Roma, physical geography needs to collaborate with underrepresented groups based on ethics and mutual respect. In one instance, Canadian physical geographers working in First Nation territory found that participatory practices, opening themselves up to distinct worldviews, and establishing reciprocity underpinned their collaboration (Kershaw et al. 2014). Physical geographers often consider themselves 'above' social position; decolonizing encourages reflection on personal and institutional positionality (Bannister 2018).

---

[7] This work is shown in a video, 'The Many Lives of a Shield' (2016), available at http://vimeo.com/200369869.

Both physical and human geographers must decolonize by learning to act appropriately in intercultural spaces and engaging in anti-racist practices. Field-based teaching holds the potential to broaden interpretations and introduce decolonial ideas. Indigenous notions of 'two-eyed seeing' are helpful as they involve 'always fine-tuning your mind into different places at once, you are always looking for another perspective and better way of doing things' (Mi'kmaq elder, in Bannister 2020: 265). Not only does this give rise to a greater depth and breadth of understanding, it also ethically encompasses previously excluded groups (Tuhiwai Smith 2012).

Decolonizing in practice: re-drawing Los Angeles

Having discussed the principles of decolonizing geography, we turn to a specific case of decolonizing in practice: a project to map Los Angeles from the perspectives of Indigenous groups. The Mapping Indigenous LA (MILA) project creates online maps to document and make visible the otherwise silenced presence of Indigenous groups in the LA area, historically (prior to European conquest, during Spanish colonialism) and today (working with Indigenous immigrant groups) (**Figure 1.2**).[8] Using text and images, MILA's maps show Indigenous displacement from territories by frontier expansion, through to the more recent arrival of Latin American and Pacific migrants. In contrast to poststructuralist geographical readings of Los Angeles (Soja 1989), the project fully engages with coloniality. Los Angeles is presented as an always-already Indigenous place at the crossroads of overlapping processes of economic dispossession, settler colonialism and transnational migration. Directly challenging dominant narratives, MILA explores Indigenous belonging and habitation in the city. By documenting Gabrielino-Tongva and Tataviam struggles as nations, MILA records centuries of Indian presence and resistance and

[8] The Mapping Indigenous LA project can be found at https://mila.ss.ucla.edu.

Figure 1.2 Indigenous Los Angeles: a plaque acknowledges Indigenous peoples and places in the city
Source: Jengod, via Wikimedia Commons: https://commons.wikimedia.org/wiki/File:Tongva_Sacred_Springs_-_Serra_Springs_-_Kuruvungna_Springs.jpg

disrupts settler expectations that 'Indians will eventually disappear'.

The Mapping Indigenous LA project demonstrates a decolonial approach to a city, Indigenous peoples, and maps. MILA draws together diverse information sources that come primarily from numerous Indigenous peoples' experiences, histories and everyday spaces. Through its 'native hub' institutional home, MILA collaborates with community groups and activists (Senier 2018). The project interprets the Los Angeles maps and photos and narratives in relation to settler colonial urbanization (**section 1.III;** Masuda et al. 2020) and Indigenous claims for rights and dignity (**Chapter 2.IV**). MILA's images and words rework standard western cartographic representations to provide glimpses of alternative space-making. By collaborating with publics, civil society and universities, MILA offers a model for decolonizing university geography. In these ways, the project decolonizes 'our [colonial-modern] knowledge of the world by extending an invitation to know it from outside the categories of western thought' (Jazeel 2011: 88).

MILA demonstrates vividly that decolonizing starts 'at home', wherever we are located. It displaces and provincializes Eurocentrism by delinking from European frames of comparison. Thus western maps take their place alongside a range of alternative cartographies, rather than replacing them. The MILA maps' text and presentation displace western ways of knowing about the world, rather than using a universal standard of what *ought* to be known. Thinking about space from the perspective of subordinated groups turns Euro-American universalism into a partial and provincial life-world and knowledge system (Chakrabarty 2000). MILA treats Indigenous communities' place-knowledges as equal to western knowledge (Mignolo 2009). As the project demonstrates, decolonizing geography brings together diverse ways of seeing and living in the world as equal but not equivalent. Los Angeles' multiple Indigenous communities and their places comprise plural histories and trajectories, re-made continuously through arrivals and dynamic relations between groups and places. This evidences shifts in ways of knowing within internal and exogenous processes, making them neither static nor relics from the past. Cumulatively, these decolonizing steps render western categories inadequate, especially the over-general terms 'Indigenous people' and 'traditional knowledge' (**Chapter 3.IV**), 'nature' and 'subjectivity' (**Chapter 4**). The MILA project carefully differentiates between named groups, placing each community in (colonial-modern) historical context.

MILA moreover challenges coloniality by entering into dignified and respectful collaborations that generate novel ways of knowing and doing (changing academic geography *and* LA's treatment of Indigenous communities). The project works closely with Indigenous neighbourhood and civil associations to gather and present the information it gathers. Doing so involves creating trust and the willingness to listen to oppressed groups' distinctive worldviews and priorities (which coloniality-modernity influences but does not determine [see **Chapter 3**]). Decolonizing, according to the Puerto Rican sociologist Ramón Grosfoguel, begins by thinking and acting collaboratively alongside and with subordinated individuals, groups and places (termed subalterns, see **Chapter 2.II**) at the margins of modernity-coloniality. Grosfoguel's approach resonates with physical and human geographies as it highlights how landscape, society and space are co-constituted in relation. 'Thinking from and with subalternized racial/ethnic/sexual spaces and bodies' (Grosfoguel 2007: 212) is a decolonial way to produce alternative connections across scales from the body, locale, region and beyond. For this reason, the geographers Jay T. Johnson and co-authors (2007) advocate creating anti-colonial geographies by embracing Indigenous knowledges and rights.

Why decolonizing geography matters

Decolonizing takes as its referent persistent and shape-shifting colonial-modernity. Decolonizing consists not merely of a changed political status or amended reading lists. It comprises ongoing struggles against and within coloniality (Holmes et al. 2014; Daigle and Ramírez 2019). The process of decolonizing does not definitively overcome and end coloniality, but occurs in the hope that its concrete pernicious effects can be challenged and lessened. Neither does decolonizing denote the *absence* of coloniality; as explained, coloniality is interwoven with modernity. But decoloniality does signal 'the ongoing ... movement towards possibilities of other modes of being, thinking, knowing, sensing and living; that is, an otherwise in plural' (Mignolo and Walsh 2018: 81). By identifying coloniality, the possibility of future alternative modernities emerges. So no fixed check-list, one-sided theory or paradigm exists for decolonizing. (For this reason, the lists and tables provided in this book are provisional, not prescriptive or definitive.) Yet via anti-colonial action the 'one-world world' (**section 1.IV**) can be transformed into what I call a 'more-than-one world', or, as the Zapatistas in southern Mexico say, a world where many worlds fit.

The aim of decolonizing is to 'open up rather than close geographies and spheres of decolonial thinking and doing'

(Mignolo and Walsh 2018: 3). Two cornerstones of decolonization can be identified: first, admitting that colonialism has occurred and is harmful; second, embarking on projects to dismantle contemporary expressions of coloniality (Vaeau and Trundle 2020). Hence, a geography fit for building and understanding a '**more-than-one-world**' needs to start with the structural transformation of its institutions, practices and relations. Too much emphasis on critical consciousness risks bypassing the urgent agendas of dismantling racist societies, restructuring land, resources and wealth, and dismantling Eurocentric world power (Esson et al. 2017). Keeping this in focus means putting the agendas and concerns of those on the colonial-modern margins at the heart of anti-colonial action (Noxolo 2017b; Daigle and Ramírez 2019). By doing so, decolonizing changes how everyone interacts with, thinks and envisions the world otherwise. A decolonial praxis goes step-by-step, acting differently, gaining grains of consciousness through new interactions, reflecting on that, and circling around to act otherwise (Mignolo and Walsh 2018: 19). Exemplifying decolonial praxis, geography collectives have formed among professional geographers, non-university scientists and communities, and civil society to produce shared knowledges about geophysical and socio-spatial issues in the Americas and Europe (e.g. Kollectiv Orangotango+ 2018).

Decolonizing broadens our conceptual and theoretical frameworks across the entire geography discipline. Readers will find discussion and examples of this throughout the following chapters. From a predominantly white, UK university perspective, a set of themes recur and are addressed in depth:

- Tackling Eurocentric white geographies through multi-racial alliances and solidarity; actively treating all with humanity and dignity in overturning colonial-modern and disciplinary practices of exclusion
- Learning from 'border' knowledges in order to reframe geography's analytical and skills toolkit, and re-conceptualizing the world from plural perspectives

- Re-analysing geophysical and social-spatial processes in relation to modernity-coloniality

Academic geography needs to get on board with decoloniality as a structural and intellectual challenge. As the particularities of decolonial priorities vary across countries, groups and institutions, geographical insights are crucial for understanding the world otherwise (Daigle and Ramírez 2019; Radcliffe and Radhuber 2020; **Chapter 3**).

VII Chapter Summary and Outline of Book

The world and its spatial relations across cities, landscapes and universities are deeply marked by coloniality. This chapter introduced the argument that coloniality is entangled with modernity: 'always already' across the 'modern' world there is coloniality. Coloniality results in acute inequalities and exclusions caused by longstanding resource flows, interpersonal relations and institutional and infrastructural orders that cumulatively result in white supremacy, racial capitalism and Eurocentric modes of understanding. For this reason, decolonizing is urgently required. Decolonizing is about everyday lived experiences; it is also about how cities interact with nature, and melting glaciers, and everything in between. It involves a profound reworking of practice *and* theory, the material *and* the mindset. In geography, decolonizing reorients practice and frameworks away from universality towards plurality and critical theorization, away from unmarked whiteness towards anti-racist practice in society.

Decolonizing praxis and thinking are gaining momentum in and unsettling geography. Responses to decolonizing imperatives include systematic and lively debates, educational programmes and new ways of doing geography. Although this book focuses primarily on English-language (Anglophone) university-level geography, due to the author's institutional and intellectual setting, decolonizing debates are happening now in the Pacific region, Brazil, the Netherlands and elsewhere, reflecting decoloniality's 'multiple and varied forms of recreating the matrix of power, knowledge and being' (Maldonado-Torres 2016: 16). Decolonizing in Aotearoa-New Zealand (where Māori geographies and territorial governance drive decolonizing debates) is not identical to decolonizing in Britain, where racism and colonial landscapes are foregrounded. Through examples from diverse contexts, the book provides insights into this variety.

Outline of the book

Chapter 2 explains the contributions of various frameworks to decolonial geographies, providing an overview of **postcolonialism, subaltern** geographies, **modernity-coloniality-decoloniality** approaches, and Indigenous and settler colonialism theories. These frameworks are relevant for human and physical geographies, as will be demonstrated in extended examples. Chapter 3 offers an overview of critical geographical work that delinks geography from its sanctioned 'one-world world' frame. Covering early decolonizing geographies, as well as Black and **Indigenous geographies**, decolonial feminisms and critical geographies of violence and peace, the chapter demonstrates that geography is alterable and able to embark on decolonizing, and argues that decolonizing means focusing on coloniality-modernity's outcomes for landscapes as spaces of geophysical and socio-spatial power and processes. Building on this foundation, Chapter 4 provides decolonizing interpretations of core concepts in human and physical geographies, and defines and exemplifies new decolonial concepts.

Decolonizing geography also transforms teaching and learning. Chapter 5 presents the reasons for and practices of decolonizing the curriculum and pedagogy. The chapter outlines core principles and discusses pathways to decolonizing physical and human geography using case studies. Since geographical research is deeply colonizing in its assumptions and practice, Chapter 6 presents a framework for decolonizing the research cycle, from design to sharing results, again with examples from physical and human geography. The chapter lays out decolonizing principles and rationales for responsible co-working, ethics and methodologies, relevant to both novice and more experienced researchers. Every chapter ends with a list of further readings and resources, including videos, blogs and websites with relevant case studies, news and discussion. Additionally, the Glossary provides brief definitions of the key terms used throughout the book.

Further Reading and Resources

Readings

Barnabas, S.B. 2015. Want to understand the decolonisation debate? Here's your reading list. *The Conversation*, 22 December, https://theconversation.com/want-to-understand-the-decolonisation-debate-heres-your-reading-list-51279.

Blaut, J.M. 1993. *The Colonizer's Model of the World*. New York, Guildford Press.

Collard, R., Dempsey, J. and Sundberg, J. 2015. A manifesto for abundant futures. *Annals of the Association of American Geographers* 105(2): 322–30.

Daigle, M. and Ramírez, M. 2019. Decolonial geographies. In Antipode Editorial Collective (ed.), *Keywords in Radical Geography: Antipode at 50*. London, Wiley, pp. 78–84.

Elliott-Cooper, A. 2017. 'Free, decolonised education': A lesson from the South African student struggle. *Area* 49(3): 332–4.

Mahtani, M. 2014. Toxic geographies: absences in critical race thought and practice in social and cultural geography. *Social and Cultural Geography* 15(4): 359–67.

Radcliffe, S.A. 2017. Decolonising geographical knowledges. *Transactions of the Institute of British Geographers* 42(3): 329–33.

Williams, M. 2016. Does your university produce racism? *Guardian*, 31 October, https://www.theguardian.com/education/2016/oct/31/does-your-university-produce-racism.

Websites

African politics reading list
http://democracyinafrica.org/wp-content/uploads/2020/08/
Decolonizing-the-Academy-The-African-Politics-
Reading-List-2020-.pdf
'This reading list is collated in solidarity with those who
are currently attempting to decolonise the university across
Africa, and beyond. ... our efforts here are simply to make
available as many sources as possible written by African
scholars.'

Black Lives Matter
https://blacklivesmatter.com
'#BlackLivesMatter was founded in 2013 in response to the
acquittal of Trayvon Martin's murderer. Black Lives Matter
Foundation, Inc is a global organization in the US, UK, and
Canada, whose mission is to eradicate white supremacy and
build local power to intervene in violence inflicted on Black
communities by the state and vigilantes. By combating and
countering acts of violence, creating space for Black imagination and innovation, and centering Black joy, we are
winning immediate improvements in our lives.'

Mapping Indigenous LA
https://mila.ss.ucla.edu
Provides a series of maps and commentaries on the Indigenous
peoples who occupied land that is now Los Angeles, and the
diverse Indigenous populations that today live and work in
the city.

–2–
Postcolonialism and Decoloniality

As discussed in Chapter 1, decolonizing geography involves delinking from Eurocentric mindsets and undoing real-world social-territorial-economic inequalities and exclusions. To build geography's capacity to address these agendas, it is helpful to draw on existing debates about living with colonialism, anti-colonial action and decolonizing our thinking. To this end, this chapter introduces four conceptual-theoretical frameworks that have influenced geography's engagement with decolonizing debates and inform ongoing decolonizing practices and thinking beyond Eurocentrism. These frameworks offer powerful tools for identifying and analysing colonial-modern relations of power, inequality and exclusion, from structural conditions to everyday realities. By combining these frameworks, geography can come to understand the structures, flows, mentalities and spatial variability of coloniality-modernity.

Recognizing that decolonizing is not guaranteed by political independence (**Chapter 1.III**), anti-colonial thinkers and leaders have identified the detrimental consequences of economic dependency, **neocolonialism** and westernization for over a century and a half, speaking from the realities around them (Young 2001; Lazarus 2004). From the 1960s, anti-colonial critiques cohered into bodies of thought in different world regions, informed by their histories of colonialism and heterogeneous intellectual currents (Bhambra 2014). From the late 1980s, these strands broadly cohered into postcolonialism, subaltern studies, modernity-coloniality-decoloniality (MCD) and interdisciplinary and Indigenous theories of settler colonialism. These anti-colonial frameworks emerged in dialogue with each other, as discussed below. They provide the analytical and theoretical bases for decolonial geographies, identifying in varying ways the material and interpretive facets of colonialism's after-effects in the present day, and generating insights into why coloniality endures and decolonizing remains a global challenge and priority. As well as differing in their methodologies and empirical foci, their interpretive bases vary considerably: postcolonialism is rooted in the humanities, whereas MCD draws more from social sciences and philosophy. In other regards, they share intellectual influences: both postcolonialism and MCD are anti-colonial and anti-imperialist, combining a moderate Marxism with critiques of racism (Leonardo 2018). These interdisciplinary frameworks are relevant for physical and human geographies, as the examples and discussion below demonstrate.

The chapter begins with postcolonialism, a disparate set of approaches unified by an ideological and political **anti-colonialism**; it is the most influential framework in human geography today. Postcolonialism provides a scholarly critique of colonial discourses and representations, giving geography the analytical tools to identify colonial powers and slippages. Subaltern geographies offer an anti-colonial framework concerned with understanding how lives dominated by coloniality are structured and lived. The modernity-coloniality-decoloniality group in turn presents geographers with an understanding of coloniality's entanglement with modernity, and the existence of plural knowledges at the borders of coloniality. The fourth strand comprises theories of settler colonialism, including approaches devised by and for Indigenous and allied scholars and movements. Each approach's key contributions to decolonial thinking are presented in this chapter, together with discussions of their limitations. The chapter concludes by suggesting that engagement with these frameworks should inform geographical theoretical-conceptual tools for decolonizing, since each framework originates in and reflects different contextual dynamics of coloniality-modernity. Hence they offer the geography discipline multiple bases for understanding colonial-modernity's spatial variegation and historic-geographical specificity, an understanding from which to interweave plural alternatives to colonial-modern living, knowing and writing.

Before starting however, a word about the terms used in the chapter. As Chapter 1 explained, decolonizing approaches question how power relations determine structures, flows and mentalities. Moreover, postcolonial and decolonial approaches argue that what we *know* is to a large extent determined by power relations. How we know about the world, what classifications we use, and the processes that confirm knowledge (e.g. scientific experiments) are, they argue, only fully comprehensible if we look at how power interrelates with knowledge-making (**Chapter 1.IV**). Postcolonial and decolonial writers use specific terms to talk about these power-knowledge relations. The term 'epistemology' refers to how we know the world and how it 'ought' to be known. For example, global science knows the age of forests by measuring tree rings. However there are multiple epistemologies in the world, some more or less dominant or institutionalized than others. The term 'subaltern' refers to social groups and their knowledges which are oppressed by existing power relations and rendered incomprehensible to dominant knowledge-holders. These terms are used through this chapter (see the Glossary for definitions used throughout the book).

I Postcolonialism: Critiques of Colonial Discourse

Postcolonialism's critique responded to widespread scholarly and public concern that discourses about, and representations of, non-western peoples and places powerfully and consistently rely on symbols, meanings and over-simplifications that originated during colonialism. Postcolonialism arises from a disparate and diverse set of theorists. Their approach is to closely examine writings and images from diverse sources – for example, development policies, western discourses about cities in the **global South**, or school maps – in order to identify the underlying presumptions and exclusions. Postcolonialism thereby contributes to an in-depth understanding of how languages, texts and images inform colonial assessments of authority, rationality and the relative importance of groups and areas around the world. Postcolonial scholars examine a range of materials – from colonial-era bureaucratic documents in archives through to present-day political and media coverage of the Middle East – to evidence the narrowness and self-justifying underpinnings of colonial power (Gregory 2004). Although initially and primarily associated with literary and historical studies, postcolonialism has proven to be a powerful tool in analysing how space is represented, organized and imagined in the present, and has informed geography's self-awareness and critique of disciplinary coloniality (Blunt and Wills 2000).

Postcolonial analysis does not take a text or picture as a neutral description of a phenomenon, but instead examines how that description is built up from selective and self-interested components. This postcolonial analysis of meanings in language and visual imagery draws on poststructuralism to reveal how colonizing and metropolitan powers produce, circulate and endorse images of places and peoples in everything from novels to social media. Edward Said's pathbreaking analysis of **Orientalism** demonstrated how France and Britain produced imaginative geographies of the Middle East that expressed European power, representation and knowledge. 'Oriental' peoples were, Said argued, represented as other, problematic or less informed in comparison to rational, powerful and knowledgeable westerners (Said 1978; Jazeel 2019) – as are Middle Eastern and Asian groups today, after 9/11 and during the Covid-19 pandemic. 'Discourse' refers to interrelations between language, practices and assumptions that give prominence to selective points of view (making them into 'common sense') and thereby obscure alternative meanings and viewpoints. In a well-known piece, the postcolonial scholar Gayatri Spivak (1999) explored how British colonial discourses regarding the practices of *sati* (Hindu widow sacrifice) selectively represented India as a problematic monolithic culture that Victorian colonial actors might improve, yet did not incorporate widows' viewpoints, which may have diverged from both colonial and (masculine) 'Indian' perspectives (Sharp 2009). Even today, politicians and others talk about 'saving' non-white women, as if patriarchy existed solely among ethnic minorities in multicultural societies, or within non-western religions.

Postcolonial geographers extend these methodologies and critiques to examine the *spatial* material and subjective consequences of colonial discourses and representations. The organization and control of space have been and continue to be integral to colonialism's material and discursive effects, as colonialism and postcolonial power transform landscapes materially and socially into 'a knowable pattern [that was and continues to be] ordered, sanitised, [and] made amenable to regulation' (Sharp 2009: 56). Across the domains of urban planning, plantations and architecture, space served and serves to organise, discipline and survey economic and political behaviours. Likewise, relationships with nature were and continue to be regulated, producing 'natural' environments that favour colonial-imperial interests and displace locals (**Chapter 4.IV**). In India for instance, colonial big-game hunting set aside tracts of land accessible to colonists and from which existing residents were displaced (Jazeel 2019).

Postcolonialism also examines the complex nature of colonial power. While that power was once thought to be all-encompassing and intractably exclusionary of colonized experience, postcolonial scholars understand that colonial relations and representations are internally contradictory. Rather than being monolithic, they are 'divided at source' (Slater 2004: 69). 'On the ground', colonization and colonialism have been less all-determining and more negotiated, resisted and reworked than metropolitans planned or expected. These power instabilities have been explained through concepts of ambivalence and hybridity (Jazeel 2019) that recognize the blurred lines between colonizer and colonized. Thus colonizers bolstered their identities by attributing opposite qualities to the colonized. As postcolonial scholars note, these binaries quickly fall apart once detailed analysis is undertaken.

Colonialism also produced psychological ambivalence, resulting in long-term impacts on social relations and political outcomes. The anti-colonial writer and psychiatrist Frantz Fanon identified these contradictions in racial divides. Under colonialism, Black people were invited to identify with white European culture and aspire to becoming European. Based on his personal experience as a Caribbean migrant in France, Fanon argued that a Black person is never allowed fully to realize this process because white racism continuously renders black bodies unrecognizable as peers (Fanon 2008 [1952]; Sharp 2009: 123). Whiteness was – and continues to be (**Chapter 1.V**) – expressed spatially, and objectifies the black body in the city, the street, schools and empires which are coded as white by default (Jazeel 2019: 129). White denial of Black subjects' dignified humanity causes psychic trauma and produces conflicted societies. The Caribbean writer Aimé Césaire (1972) argued that everyday colonialism is pathological for colonizers and colonized alike. Fanon's conclusion that racial exclusion was an enduring structural factor in colonialism and postcolonial histories was a highly influential contribution to anti-colonial theorizing (**section 2.III**, and **Chapter 3.IV**).

In the early 1990s, the postcolonial scholar Homi Bhabha revisited the question of postcolonial ambivalence and argued that a colonized subject expresses both resistance to, and complicity with, colonial power (Bhabha 2008). Taking a long view of colonialism and its ongoing denial of full recognition to racialized and colonized groups, Bhabha saw postcolonial societies – whether in the global South or in multicultural western cities – as characterized by dynamic unpredictable encounters between cultural forms, meanings and identities (Bhabha 1994). Whereas colonial power attempted to 'fix' a stable culture, postcolonial subjects can be, he argued, ambivalently in-between, subverting or disrupting colonial hierarchies. Critiques of Bhabha focused on his positive spin on hybridity, which downplayed the enduring hierarchical structures that determine relative status in postcolonial societies, yet his key point about the fluidity and borrowing between and across cultures and identities has continued to be important.

Moreover, Bhabha's analysis of ambivalence and hybridity reminds us to query any claims about unchanging traditions or social divides, and to closely examine grounded and 'messy' social identities. When postcolonial work focuses on the present, its ethnographic methods shed light on the daily practices and spaces of colonial power and the lived experiences of marginalized and racialized populations (de Leeuw and Hunt 2018: 4). These accounts document how individuals and groups experience dehumanizing and dispossessing interactions and institutions viscerally and in everyday interactions, making them by turns invisible or hyper-visible. The 'connectivities joining colonial pasts to ... presents' thereby persist, although the current 'geopolitical and spatial distribution of inequalities are not simply mimetic versions of earlier imperial incarnations but refashioned and sometimes opaque and oblique reworkings of them' (Stoler 2016: 4–5). Under postcolonial modernity, life is 'a struggle to make it from today to tomorrow' (Mbembe 2010: 673) due to structural conditions that impose conditions of temporariness and precarity on marginal groups.

Through a critical dissection of colonial narratives, postcolonialism furthered an understanding of how knowledge is made and circulated. Edward Said's account of Orientalism highlighted how knowledge and power worked together, the west's information-gathering comprising an expression of and a reassertion of power. The anthropologist Fernando Coronil detailed how **Occidentalism** – the 'forms of classification, hierarchy, exclusion, naturalization and spatiality rooted in the deployment of global power' – is the obverse of Orientalism (Coronil 1996: 53). Occidentalism is profoundly geographical, separating the world into bounded units, naturalizing them and reproducing asymmetrical geopolitical power relations (Slater 2004). Another facet of colonial power-knowledge relations was (and is) the **epistemic violence** 'done to the ways of knowing and understanding of non-western [subalterns including] indigenous peoples' (Sharp 2009: 111). Colonial power silenced and disarticulated other knowledges and thereby reinforced its self-identification as the bearer of a superior and universal knowledge. Postcolonial theories have implications for physical geographical approaches to geophysical processes and features, as they identify the power of enduring Eurocentric images of primordial non-western landscapes (including glaciers, Australian dry areas, or equatorial forests). Postcolonial theory identifies these 'fixed' representations and provides a means to critique and question them.

Limitations of postcolonialism

Despite postcolonialism's powerful analytical lens, it has been critiqued on a number of grounds which open up further issues (see later sections). In this section, issues of western-centrism, coverage, and text versus lived experience are discussed. First, whereas mid-twentieth-century anti-colonialism highlighted economy, politics and race, late twentieth-century postcolonialism was inspired more by literary and historical approaches (Lazarus 2004). Although critical, postcolonialism tended to focus largely on metropolitan-colonial self-referential worldviews and hegemonic cultures. For instance, Said's book *Orientalism* was criticized for homogenizing the west, emphasizing continuities over discontinuities, not questioning the Orient's gendered associations, and working solely with texts (Sharp 2009: 27–8). Colonial discourses *are* still present and widespread today but, critics argue, how does pointing that out help dismantle colonial power? Political domination and economic exploitation, structures of dependency and neocolonialism, remain relatively neglected topics in postcolonialism. Nevertheless, postcolonialism's literary and historical focus on subtle historical processes and the intricacies of language avoids reducing coloniality to a rigid, all-explaining process, while recent postcolonialism has highlighted unpredictable temporalities and differentiated patterns of postcolonial duress (Stoler 2016).

Second, postcolonialism focused primarily on the historical and geographical contexts of South Asia and sub-Saharan Africa, especially in relation to northwest European colonialisms, using European social theory. Postcolonialism's brilliant exponents – Homi Bhabha, Gayatri Spivak and Edward Said – had personal histories of postcolonial displacement and commitments to excluded groups, but as postcolonialism became institutionalized in Anglophone poststructural theory and Euro-American universities, it tended 'to invalidate or diminish the significance of reflections on colonialism developed from other locations and perspectives' (Coronil 2004: 237; Lee and Cho 2012). Postcolonial insights were applied to multicultural immigrant societies, but what about the settlers displacing Indigenous groups? Postcolonialism addressed hierarchies of caste and tribe, yet did not examine the colonial formations of **indigeneity** and enslavement.

II Subaltern Studies

Subaltern studies focuses on subordinated groups and individuals who are misinterpreted, neglected, oppressed or marginalized by prevailing power relations, especially in colonial and colonial-modern contexts. The subaltern studies approach is associated with the Indian scholarly collective known as the Subaltern Studies Group (SSG), which came to prominence in the 1980s. In contrast to postcolonialism, the SSG was avowedly Marxian in approach and sought to examine history from 'below', documenting the histories and agency of social groups and areas unaccounted for in colonial discourse and in postcolonial histories and cultures. In focusing on people's power, history and voices, the SSG significantly influenced how we understand coloniality today (Spivak 1988). Sharing the Indian group's interests, a Latin

American debate around subaltern histories and experiences emerged in the early 1990s (Coronil 2004).

Why were these scholars interested in subaltern groups? Like the postcolonial approach, they focused on how dominant power perpetuated itself, in part through knowledge and representation. Additionally, however, they placed their insights more firmly in relation to institutional power and explored the capacities of non-dominant groups for alternative modes of acting and thinking. To assist this analysis, the SSG drew on earlier Marxist discussions around power and contestations. Specifically, they modified the Italian thinker Antonio Gramsci's (1891–1937) concepts of hegemony and the subaltern. Hegemony refers to the cultural and institutional power to provide persuasive accounts of the world that convince societies to believe and to act in certain ways. Thus, Gramsci argued that the Italian Catholic Church, large landowners and the bourgeoisie used ideology rather than force to become hegemonic, despite widespread poverty and exclusion. Subaltern studies writers saw parallels between Italy and postcolonial India, where bourgeois political and cultural hegemony excluded the vast majority of the population after political independence. In this and other contexts, subaltern approaches suggest that excluded sectors are subaltern, meaning their perspectives do not inform the dominant culture, identity and decision-making (Jazeel 2019).

Subaltern, then, is a **relational** concept, a position understood through relationships with hegemonic institutions. Hegemony refers to the power of establishing what is common sense, the status quo, or 'correct' in a specific situation. For Gramsci and the Subaltern Studies Group, hegemony is always incomplete, a work in process, because it relies on continuous adjustments and inputs (through media, speeches, education and so on) to render the status quo normal and unquestioned. Another factor behind the incompleteness of hegemony is the autonomous emergence of critique and alternative interpretations (sometimes in cultural institutions such as labour unions and social movements) in a counter-hegemonic action. The situation of the subaltern thus depends on context and on the relations between coloniality and resistance, with subalterns being neither fully integrated into nor completely separate from the dominant society. Not everyone with alternative views or demands for change is a subaltern (Sharp 2009).

The concept of subaltern is useful in understanding social sectors and individuals whose lived experiences and systems of thought render them subject to forms of coloniality. In terms of positionality and social identity, however, the term subaltern encompasses highly varied groups with heterogeneous, cross-cutting divisions and differences. For example, undocumented migrants attempting to reach Europe in precarious boats are subaltern as they are rendered powerless by international law, yet they are extremely diverse in terms of generation, nationality, gender, ethno-linguistic grouping, and capabilities. Subaltern debates document the diversity of marginalized populations in order to avoid colonial-modern representations of timeless cultures. 'All subaltern communities are internally differentiated and conflictual … [forging] political unity or consensus in painfully contingent ways' (Florencia Mallon, in Coronil 2004: 232). Subalterns are diverse and tensioned precisely because they are contingently located within *and* outside modernity and coloniality, such that 'no essential subordinated subject [exists] entirely separate from the dominant discourse' (Blunt and Wills 2000: 192).

On the basis of these concepts, subaltern studies writers including geographers raise questions about how it is possible to know about subalterns. Their epistemological concern is to identify the power relationships between subalterns and those who write about, analyse or represent them, whether in a colonial document or in a scholar's 'history from below'. Under hegemonic colonial discourses and epistemic violence, it is already difficult to discern the traces of subaltern thinking and experience. Gayatri Spivak doubted that a subaltern voice could be heard in the archives, as colonists wrote and recorded what was relevant for themselves (Spivak 1988). However, in their project to rewrite Indian history from marginalized actors' perspectives, the SSG group closely analysed colonial archives to read 'against the grain' and found evidence of subaltern actions and thought. Through this work, subaltern studies raise important questions about whether entities such as the state or the nation always reflect the models used in western contexts and western theory (Jazeel 2019). However, subaltern voices can be essentialized, oversimplified and romanticized in social science methods and research (Sharp 2009). To avoid this, scholars pay attention to the slippages and ambivalences in colonial-modern identities. Subalterns often occupy awkward in-between positionalities. For example, British Empire employees in India were encouraged to mimic European outlooks and behaviours, yet colonial society would never recognize them as fully British (Bhabha 1994).

In other situations, subalterns deliberately occupy epistemic borders in order to better contest and subvert dominant representations. Chicana feminist Gloria Anzaldúa situated herself outside the parameters of US and Mexican nationalisms and sexual norms, refusing a privileged position and disrupting the dominant social categories available to her (**Box 2.1**). Anzaldúa highlights an important dimension of subaltern positionality, namely the relation between subalterns and postcolonial nationalism. Anti-colonial struggles for independence rely on narratives of belonging to galvanize support in newly formed nations (**Chapter 1.III**). Yet postcolonial nationalisms are frequently constructed on the basis of elite and power-holders' exclusionary cultural values and social positions, and expressed through institutions that meld local cultural forms with colonial models, as occurred in India (Chatterjee 1993). Elsewhere, indirect colonial rule institutionalized fluid pre-conquest ethnic and religious identities; these categories in turn shape expressions of postcolonial nationalism, with new forms of oppression and belonging. Throughout the twentieth century, postcolonial nationalisms interacted with ideologies of development and progress to transform the political-economic realities and cultural representation of subaltern groups (Fanon 2004). Postcolonial nationalism thus contributes to subalterns' heterogeneity, even as it reinstates elites as hegemonic economic, social and political actors.

Box 2.1 Overlapping borders and identities

Gloria Anzaldúa (1942–2004) was a writer and poet whose personal experiences led her to unpack the complex relations between and across postcolonial spaces and social identities. Occupying in person and analytically the contradictory zone of the US–Mexico border, Anzaldúa celebrated the 'mixed consciousness' (*conciencia mestiza*) held by herself and others – a consciousness, she argued, that reflected in-between and overlapping racial, gendered and geopolitical relations.

> The actual physical borderland that I'm dealing with ... is the Texan-US Southwest/Mexican border. The psychological borderlands, the sexual borderlands and the spiritual borderlands are not specific to the Southwest. ... It's not a comfortable territory to live in, this space of contradictions ..., [marked by] hatred, exploitation. However there are compensations ... keeping intact one's shifting and multiple identity and integrity. ... The US–Mexico border *es una herida abierta*, where the Third World grates against the first and bleeds ... the lifeblood of two worlds merging to form a third country – a border culture. ... My Chicana identity is grounded in the Indian woman's history of resistance ... I feel perfectly free to rebel and to rail against my culture. ... What I want is an accounting with all three cultures – white, Mexican, Indian. ... And if going home is denied me, then I will have to stand and claim my space, making a new culture – with my own lumber, my own bricks and mortar, and my own feminist architecture. (Anzaldúa 1987: Preface, 3, 21)

Anzaldúa constructs a politics of knowledge at the international frontier, and a 'body-politics' of knowledge as a Chicana lesbian body.

In summary, the concept of subaltern focuses on social
and knowledge relations between individuals and groups
holding persuasive cultural and institutional power (such
as colonial officials, state bureaucrats, journalists and
progressive scholars) and the individuals and groups they
seek to influence and 'speak for'. Despite varying perspectives, it is agreed that subalterns experience modernity from
its margins, and that their knowledges and experiences
are key to anti-colonial critiques. In broad comparison
with postcolonial work, subaltern scholarship focuses more
tightly on the power-knowledge relations between dominant
and subordinate actors, seeking a way to bypass the power
inherent in academic writing and to give space to oppressed
groups. Geographers contribute to these debates by mapping
places and geographies where subalterns exert agency, despite
hegemonic space-making and spatial epistemologies (Jazeel
and Legg 2019). Physical and environmental geographies can
take lessons from subaltern approaches, as they foreground
groups and knowledges that disrupt prevailing common
sense and offer alternative insights around water, plants,
glaciers, coasts and atmospheres (**Box 2.3**).

Limitations of the subaltern studies approach

The subaltern studies approach has been critiqued for a
number of limitations. Firstly, the term itself is slippery,
as scholars use it in different ways, sometimes to describe
a situation of oppression, and at other times theoretically
(Jazeel and Legg 2019). A related issue is that subaltern
studies easily falls into binaries – elites versus subalterns,
subalterns versus the postcolonial state – that oversimplify
the overlapping relations of influence and the ambiguities
of hyper-diverse societies and **intersectional** positionalities.
Critiques also focus on what the geographer Jovan Scott Lewis
terms the 'necessary persistence of the colonial referent' (Scott
Lewis 2018: 26), whereby subaltern scholars refer constantly
to how colonizers see the world. To break from colonial
mindsets means thinking more broadly about the overlapping
dynamics of modern-day coloniality. Scott Lewis (2018)

suggests using transdisciplinary and diaspora approaches to engage in comparative, heterogeneous, spatially attuned analysis. On a related point, subaltern studies is criticized for relying on European theorists (especially Michel Foucault) and ignoring vibrant Southern theory (Connell 2007), and for sidelining class, labour and gender dynamics. Such lacunas contribute to subaltern studies' complex yet attenuated relations with wider political struggles.

III Modernity-Coloniality-Decoloniality (MCD) Group

Although geographers have only recently engaged with the modernity-coloniality-decoloniality approach, the MCD group had begun Spanish-language discussions between sociologists, anthropologists, cultural theorists and philosophers in the 1970s, focusing on diverse anti-colonial and postcolonial literatures in Latin America and the Caribbean. After translations began to be produced, Anglophone geographical discussions around decoloniality increasingly engaged with MCD understandings of coloniality and decoloniality. Comprising neither a paradigm nor a theory, the MCD framework differs from postcolonial and subaltern arguments in its characterization of modernity, coloniality's time-frame and extent, and **knowledge production.**

The central MCD argument is that modernity and coloniality are inextricably interlocked, so to talk about the modern world necessarily involves acknowledging and understanding coloniality. Yet coloniality-modernity also contains within it the possibility of transformation – decoloniality; hence the MCD label refers to processes that both divide *and* connect. Modernity has a dark side – coloniality – that originated in the late fifteenth century, when southern European countries led a world-changing transformation (**Chapter 1.III**).[1] Under

---

[1] By comparison, postcolonial and subaltern frameworks emphasize the post-Enlightenment period of northern European capitalist expansion and colonialism. This responds to MCD critiques of postcolonialism for its limited poststructuralist analysis of power.

the monarchy and Catholic Church, the Spanish re-conquered
the Iberian peninsula after over 400 years of North African-
Moorish governance, culminating in Granada's capture in
1492. In that same year, the Catholic Kings issued Christopher
Columbus with a unique Latin-language 'passport' permitting
him to traverse the Atlantic in an attempt to reach India.
Meanwhile, Portuguese ships had begun circum-African
voyages to Asia, followed by entry into what became Portugal's
colony of Brazil. Spain and Portugal inaugurated systems of
plunder and violence that transformed the incipient world
trade-manufacturing system into an increasingly European-
dominated one (Abu-Lughod 1989). Conquest of the Americas
– consolidated with Indigenous and Black enslavement and
land capture – initiated coloniality-modernity as an **Othering**
of non-Europeans, a capitalist world system and a universal
system of knowledge. In combination, these relations of
power, knowledge and society reoriented the world spatially
from the Mediterranean to the Atlantic:

> The conquest and colonisation of America is the
> formative moment in the creation of Europe's Other; the
> point of origin of the capitalist world system, enabled
> by the gold and silver from America; the origin of
> Europe's own concept of modernity; the initiation point
> of Occidentalism as the overarching imaginary and self-
> definition of the modern/colonial world system (which
> subalternized peripheral knowledge and created, in the
> eighteenth century, Orientalism as Other). (Escobar
> 2007: 204)

As this quotation suggests, MCD came to its understanding
of coloniality through dialogues with postcolonialism and
subaltern studies (Leonardo 2018).

According to MCD scholars, modernity is a complex,
spatially differentiated pattern of relations, starting with
Spain's and Portugal's conquests and control of west
Africa's coastline and of South America. That early form
of colonial-modernity was later superseded by northern
Europe's Enlightenment science, philosophy and politics

and its distinct forms of colonialism. Later still, North America provided a template for political and cultural modernity, involving imperial interests across the globe (Slater 2004). Over successive centuries, the MCD group suggests, modernity was constituted within the colonial matrix of power, a globally hegemonic formation since 1492 that articulates race, labour, space and subjects to the benefit of capitalism and Europeans (and their settler descendants). In this framework, modernity is, at the same time and through the same processes, coloniality; the two are inseparable sides of the same coin. Modernity-coloniality refers to the heterogeneous ensemble of processes and social formations that encompass modern colonialism and colonial-modernities, with globe-spanning yet geographically and historically differentiated spheres of power, knowledge and being (Escobar 2007).

In the MCD argument, coloniality does not solely reside in discourse and representation but integrates economics, politics, knowledge and social relations. The colonial-modern world system thus comprises a structurally heterogeneous ensemble of processes and social formations (Escobar 2007). The Peruvian sociologist Aníbal Quijano (1928–2018) inter-linked political-economic, cultural and epistemic processes in his account of coloniality. Four mutually articulated domains underpin coloniality: expropriation of land and exploitation of labour; control of authority (military and state forms); control of gender and sexuality (the Christian family and sexual norms); and control of subjectivity and knowledge. These together constitute the colonial matrix of power (Wynter 2003; Mignolo 2007). Trans-Atlantic capital circulation and forced Indigenous and Black labour forged an inherently exploitative racial capitalism (Grosfoguel 2007; Lugones 2010). MCD stresses the dynamic nature of coloniality-modernity, which continuously re-colonizes on new bases generating new forms of exclusion through transformed spiritual, racial, gender, knowledge and sexual hierarchies (Quijano 2007; Rivera Cusicanqui 2012).

In comparison with postcolonialism and subaltern studies, MCD scholars consider racial difference to be central to

coloniality. They trace the ideology of biological inferi-
ority and superiority from the Spanish re-conquest of
Moorish territory and the 'New World' hierarchies between
Europeans, colonized Indigenous, enslaved Black Africans
and non-Christian Europeans (Wynter 2003). Racialization
underpinned the organization of unequal control over land,
labour, resources and knowledge, and the dispossessing and
displacing of Black and Indigenous populations assigned
to distinct sectors and spaces (**Chapter 1.V**). Race became
intertwined with capital and Eurocentric knowledge, generating globe-spanning universal classifications (Quijano 2000).
Racialization in this framework was forged at the same time as
capitalism, and does not precede it. Race, moreover, is interlocked with hierarchical relations of gender and sexuality,
as coloniality's social classifications legitimized the interconnected processes of exploitation and domination. This matrix
of power controlled work, nature, reproduction, sex, subjectivity and authority, in what MCD terms the coloniality of
being (Lugones 2007; Maldonado-Torres 2007; **Chapter 3.IV**).
These dynamics in turn informed the distinction between valid
knowledges (white, western) and other knowledge systems,
which were suppressed (de Sousa Santos 2014; **Chapter
1.IV**). MCD writers identify these colonial-modern forms
of racialization, rationality and power across present-day
government, education and social relations. In this sense,
coloniality pervades everyday life, although MCD writings
focus primarily on global, century-long dynamics.

Echoing subaltern studies in some respects, the MCD
approach finds in subaltern groups a range of forms of agency,
knowledge and subjectivity that operate within and against
coloniality. So coloniality is not all-encompassing (Mignolo
and Walsh 2018). Despite hegemonic Eurocentrism, subalterns live everyday lives and hold knowledges formed at the
colonial-modern margins (see example in **Chapter 1.VI**). This
generates 'thinking from another place, imagining an other
language, arguing from another logic' (Mignolo 2000: 313),
a thinking that is irreducibly heterogeneous because of plural
colonial-modern realities. In the US context, J.D. Saldívar
describes these other logics:

Border thinking ... emerges from critical reflections of (undocumented) immigrants, migrants, *bracero/a* workers, refugees, *campesinos*, women, and children on the major structures of dominance and subordination of our times. ... Border thinking is the name for a new geopolitically located thinking of epistemology from both the internal and external borders of the modern (colonial) world system. (Saldívar quoted in Wright 2019: 512)

Critical of Occidentalism and the exclusion of non-western thought traditions, border thinking's impure logic reflects contradictory positions within and against coloniality-modernity (Escobar 2007: 205). Exemplified by Gloria Anzaldúa's *conciencia mestiza* (**section 2.II**) and Andean notions of life in plenitude or *sumak kawsay* (Radcliffe 2012), border thinking holds the potential of decolonizing (Mignolo and Walsh 2018).[2] Unlike the western 'place-less', ego-centric knowledge, border thinking is located in relation to both geopolitics and intersectional social relations (Mignolo 2002; **Box 2.2**). Border thinking is understood to challenge the Eurocentric **geopolitics of knowledge production**.

In comparison with postcolonial and subaltern studies approaches, Black feminists and the MCD school view the subaltern as capable of autonomous knowledge, outside the bounds of hegemonic institutions and discourses. Subalterns' distinct non-mainstream knowledges, they argue, are rooted in lived experiences of and critical reflections on the 'dark side' of modernity. The African-American writer bell hooks highlights the vibrancy of this geopolitical and intersectional position: 'Space in the margin ... is a site of creativity and power, that inclusive space where we recover ourselves, where we move in solidarity to erase the category of colonized/colonizer' (hooks quoted in Sharp 2009: 115).

---

[2] MCD analysis of border knowledges rests on work by Frantz Fanon and Gloria Anzaldúa, in place of postcolonial references to Michel Foucault and Thomas Kuhn (Mignolo 2007).

Box 2.2 Coloniality and the who/where of knowledge

The knowledgeable subject is a person or group who learns and then embodies knowledge about the world. European understandings of the knowledgeable subject have historically drawn on the French philosopher René Descartes' (1596–1650) famous dictum, 'I think therefore I am' (*cogito ergo sum* in Latin). By this he meant that knowledge could be accumulated by a knowledgeable individual through objective, disembodied and neutral work to address human earthly concerns. Feminist scholars convincingly argue that (Eurocentric) scientific claims to adopt a 'gaze from nowhere' in fact obscure and universalize particular standpoints (Haraway 1988). Feminist geographers and MCD scholars show how disciplinary interpretations of the world presume masculine gendered viewpoints, which sideline and devalue other embodiments, such as women, **BIPOC** and the differently-abled.

MCD scholars – led by the philosopher Enrique Dussel – suggest that any modern knowledgeable subject is irredeemably entangled in coloniality. Descartes had pronounced knowledge to be universal, and held by an individual subject. However, Dussel argues, for a century and a half *before* Descartes, Europeans had been gathering information and formulating knowledge claims because of and in the context of modernity-coloniality. Hence Descartes' knowledgeable subject was already a *conquering* subject, whose knowledge is set within this violent and exclusionary project: 'I *conquer* therefore I am' (Maldonado-Torres 2007).

English-language geography – among others – is not immune to these colonial-modern contradictions. It endorses a disciplinary imaginary that it knows about the world, yet its research and teaching reflect the worldly relations of coloniality-modernity (Raghuram et al. 2009; Abbott 2006).

Owing much to bell hooks, the MCD approach sees subalterns' experiential knowledge as the basis for alternative epistemologies that disrupt colonial-modern universalism (**Chapter 1.IV**). 'Modernity-coloniality's underside, margins, and cracks' become the site for **border knowledges** as they encompass 'a geo- and body-politics of knowledge that reveals both the racial and gender foundation of white ... hegemonic epistemology' (Mignolo 2007: 485; Mignolo and Walsh 2018).[3] For this reason, Eurocentric knowledge systems cannot represent or easily understand subaltern knowledge systems.

In summary, decoloniality implies that social and knowledge relations need to be fundamentally reorganized. Building on the rejection of both (Euro-American) modernity and coloniality, decoloniality offers plural possibilities of exiting from coloniality-modernity (Mignolo and Walsh 2018).

Limitations of the MCD approach

The MCD approach reflects several limitations. Its broad historical sweep and use of abstract concepts such as capitalism, knowledge and being bring charges that MCD does not follow up its ambitious theoretical framework with detailed, contextualized, empirical work. Human geographers working in feminist, political and development geography have begun qualitative research to explore local contexts where coloniality and decolonizing actions express place-specific meanings and processes (Radcliffe 2015, 2019). Such work addresses another criticism of MCD, namely that, despite Aníbal Quijano's inclusion of gendered power as an integral element in the colonial matrix of power, the MCD group engages unevenly with feminisms (Lugones 2007, 2010; **Chapter 3.IV**). Further, although the MCD approach works on behalf of border thinking, critics find that its dense theoretical language and limited discussion of its politics of knowledge constrain its applicability and sensitivity (Asher 2013; compare Escobar 2008). With its primary focus on

---

[3] Decolonial feminisms develop Spivak's close attention to the politics of scholarship and voice (**Chapter 3.IV, Chapter 4**).

knowledge and epistemic liberation, MCD is also a target for scholars and activists who emphasize decolonial objectives of directly – and materially – achieving territorial claims and anti-racism goals.

In summary, MCD scholars argue that decoloniality is a process 'born in responses to the promises of modernity and the realities of coloniality ... which undoes, disobeys and delinks from modernity-coloniality' to construct alternative modernities (Mignolo and Walsh 2018: 4, 7). Despite its limitations, MCD has usefully advanced discussions about what decolonizing means in the modern-colonial world, as an 'ongoing serpentine movement towards possibilities of other modes of being, thinking, knowing, seeing and living; that is, an otherwise in plural' (Mignolo and Walsh 2018: 81). For physical and environmental geographies, MCD prompts critical reflection on scientific approaches and provides the theoretical bases for engaging with alternative 'border' knowledges (**Box 2.3**).

IV Indigenous and Settler Colonialism Theories

Settler colonialism refers to a form of colonialism whereby incoming – primarily European – colonizers come to settle on colonized lands and dispossess the previous inhabitants (**Chapter 1.III**). Decolonial critiques range from interdisciplinary work on settler colonialism as a structure of propertied, legally mandated power, to diverse Indigenous theories that track settler colonialism's destructive consequences for Indigenous worldviews, societies and territorial control. This section outlines the frameworks and key features of geographies of settler colonialism, including Indigenous arguments and proposals.

Indigenous and settler colonialism critiques identify the processes that lead incoming settlers to displace resident (Indigenous, Aboriginal, 'native' or other) populations from their historic territories and associated life-worlds (Wolfe 2006). These processes comprise an unfinished conquest, incorporation into racial capitalism, and violence (towards, e.g., Russian peasants, or Latin American and US Indigenous populations). Settler colonialism comprises a structure – not a single moment of conquest – and establishes patterns of violence, social restructuring and knowledge destruction that continue beyond initial occupation, often for centuries (de Leeuw and Hunt 2018; Stoler 2016), as internal colonialism. Originating in European projects, various settler colonial forms are active in North America, southern Latin America, Israel-Palestine, Australia and Aotearoa-New Zealand (on a US case, see **Chapter 4.I**). European settler colonialism had profound impacts too in Ireland and across Africa (Cavanagh and Veracini 2016; Gordon and Ram 2016). Across these areas, settler colonial geographies exhibit three core features:

• Settlers come to stay in colonized lands, intending to occupy them
• Settler colonial structures of law, education and governance seek to eliminate Indigenous/resident groups, and assert control over their lands
• Racist narratives dehumanize residents; settlers self-identify as founders/originators of the country

Settler colonialism draws on colonial-modern spatial notions and practices. The western legal category of 'empty lands' (*terra nullius*, see **Chapter 4.I**) functioned to deny prior habitation, represented native populations as obstacles, and legitimated settler seizure of land for 'productive' agriculture or cattle-ranching.

Over time, a settler entitlement logic becomes institutionalized in relation to the possession of and rights over land, and the settlers' cultural and demographic reproduction (Velednitsky et al. 2020). Settler priorities determine the limited extent or withholding altogether of prior residents' territorial control (Sidaway et al. 2014; Howard-Wagner et al. 2018; Kedar et al. 2018). Settler goals necessitate the destruction of native society and its replacement, practised through forced religious conversion, the forced placement of Indigenous children in boarding schools and the removal of native land title. These practices systematically undermine the existing inhabitants' socio-territorial-knowledge relations, creating spatial dynamics that transcend public/private and local/regional spaces (Banivanua Mar and Edmonds 2010). Geographers examine the biopolitical and gendered-raced dynamics of settler colonialism, demonstrating how diverse spaces and domains, from urban infrastructures to spaces in the home, are constituted through these unstated yet pervasive logics (Naylor et al. 2018). In mid-2021, Canadian First Nations again faced evidence of settler violence with the discovery of thousands of unmarked children's graves in former residential schools. Between the mid-nineteenth century and the 1970s, religious and state residential schools forcibly removed Indigenous children from communities, forbade their languages and treated them as prisoners and cheap labour (de Leeuw 2017a). Across the Americas and Australasia, Indigenous life expectancy and infant mortality are consistently worse than in non-Indigenous populations. In settler societies, gender violence is more forceful against Indigenous women and girls than non-Indigenous (de Leeuw 2016; **Box 4.2**).

In the face of settler colonial onslaughts, Indigenous activists and scholars devise distinctive and plural practices and theories to analyse and bypass settler geographies and powers. Against the denial of territorial rights, Indigenous theorists and movements articulate Other forms of **sovereignty** and political entities, refusing nation-state jurisdictions and multicultural policies (A. Simpson 2014; Coulthard 2014). From diverse perspectives, they seek to achieve Indian resurgence and decolonization through Indigenous practices of cultural, political and spiritual revitalization. This politics of resurgence draws on anti-colonial, land-based learning, agency and epistemologies to guarantee survivance (an Indigenous concept that means survival and endurance) (Daigle 2016; Simpson 2011; Nirmal 2016; Corntassel 2020; **Chapter 3.IV**). Indigenous and settler colonialism theories bring to physical and environmental geographies an awareness of ongoing colonialism in many Anglophone contexts, and of knowledge systems operating against unequal power relations (**Box 2.3**).

Box 2.3 Connecting physical geography and postcolonial-decolonial approaches: weather, climate and Aboriginal knowledges

Weather is increasingly appreciated as a phenomenon directly and significantly impacted by anthropogenic activities. Moreover, weather is deeply shaped by colonial-modern dynamics, as an Australian case demonstrates. In order to decide on the location of a capital city, British settler colonizers in Australia took into account the continent's weather and climate patterns. Parliamentary debates referred to 'scientific' interconnections between weather, race and colonialism. Specifically, MPs spoke about the importance of locating the capital in a temperate zone, best suited to white European governance and power – arguments that presumed their right to rule. This information and the institutional context ensured the 'colonists actively invisibilised the fact that diverse weather-places were already supporting human and more-than-human systems of governance and legal orders' (Wright et al. 2021: 209). Prior relations to place-specific weather were ignored, and Aboriginal presence in what became Canberra was interpreted through deeply racialized mind-sets. Today, Aboriginal communities and other-than-human weather entities (including storm cloud Milpirri) – collectively known as Country – make their home in Canberra, despite ongoing settler colonialism over daily and generational time scales (Wright et al. 2021). This case illustrates the importance of understanding coloniality-modernity in relation to physical geography, in order to better address locally expressed and profoundly significant processes that underlie and influence geophysical-hydraulic-fluvial processes.

The case also illustrates how the frameworks outlined in this chapter enrich physical geographical understanding. Postcolonial approaches examine colonial discourses and representations to pinpoint colonial mindsets and imaginative geographies, as in the British parliamentary discourses about Australian patterns of racial difference, weather and colonial governance. Subaltern studies seeks to identify relations of power that exclude knowledges. In colonial Australia, the British authorities ignored Aboriginal understandings, instead drawing on highly racist interpretations of evolution. MCD identifies systematic coloniality in modernity, and draws on oppressed groups' knowledges and critiques. In Australia today, Aboriginal scholars and groups challenge their marginalization and disadvantage, at the same time developing knowledges and practices that work with weather features to ensure survivance of Indigenous peoples *and* Country. These practices and knowledges exemplify Indigenous theoretical discussions about resurgence and land-based learning. Aboriginal experiences in Australia are deeply rooted in the power relations, social relations and knowledges forcibly established under settler colonialism, which continues today (Howitt 2020). As this example illustrates, physical geographical topics – weather and climate in this case – can engage and contribute to decolonizing geography.

Limitations of Indigenous and settler colonialism theories

Settler colonial and Indigenous theories have been criticized on a number of points. First, the settler colonialism framework risks reducing socio-spatial dynamics to a settler–colonized binary. In reality, settlement and occupation are complex and dynamic. Settler colonialism in the Americas relied from the start on Black enslavement, resulting in a 'triad of relations between settler-native-slave' (Tuck and Yang 2012: 17). Over time, the incorporation of indentured and diverse immigrant labourers further complicated intergroup relations, prompting critical reflection on anti-Blackness and anti-Latinx aspects (Daigle and Ramírez

2019; Tuhiwai Smith et al. 2018). Anti-Blackness does not derive from settler colonialism and has ramifications for anti-colonial practice and decolonial theorizing in settler colonial societies (Wynter 2003; Garba and Sorentino 2020). Moreover, human relations with multiple non-human beings and entities demonstrate the complex interweaving between social sectors in settler colonies (Braverman 2021).

Critical debates also focus on the directions and forms of decolonization, especially in societies with highly diverse settler and immigrant populations, and their resulting divergent positions (Pulido 2018). While the logic of settler colonial theory points to settlers' removal and the return of territory to Indigenous and prior residents (Tuck and Yang 2012), the practical demands of Latin American Indigenous groups are for political and cultural autonomy. In these and other plural settler societies, the politics of change highlights the urgency of unsettling unspoken settler privilege and learning to listen to Indigenous knowledges, in order to facilitate equitable coexistence (Howitt 2020).

V Chapter Summary

Chapter 1 established that colonial structures are pervasive and globe-spanning (while being geographically varied), making the decolonizing of geography and the world it describes both urgent and necessary. Building on those points, Chapter 2 has examined the conceptual-theoretical frameworks that describe, critique and seek to undo coloniality. In varied conceptual and analytical terms, the frameworks of postcolonialism, subaltern studies, MCD and settler colonial and Indigenous theories offer significant tools for geography's task of decolonizing, as they identify the key underlying processes and elements of coloniality-modernity (**Table 2.1**). Geography moreover benefits from these frameworks' origins in and commitment to majority world peoples and places across physical and human geography (**Box 2.3**). Always theorizing in relation to colonialism and imperialism, the frameworks introduced here exemplify what the sociologist

Raewyn Connell calls Southern Theory (Connell 2007), that
is, theory that radically reworks Eurocentric social theory.
Together they ensure that decolonizing makes a 'louder and
more radical challenge ... and [permits] direct confrontations
with existing practice' (Noxolo 2017a: 342).

Bringing these frameworks together highlights four points:

- Each framework provides critical insights into colonial
  power, knowledge and subjectivity
- Each framework has varying relevance to different world
  regions, responding to coloniality's variegated and uneven
  expressions
- In this sense, the frameworks are not directly comparable
  due to their distinctive origins, scholarly approaches and
  conceptual bases
- Debates around geographical decolonizing benefit from
  these approaches, separately and in dialogue

The four approaches together remake colonial-modern
geography and assist it in delinking from universalism.
The pluralizing of geographical conceptual-theoretical
frameworks is intrinsically and constructively decolonizing.
Despite the differences and tensions between them, the
frameworks contribute to the discipline's decoloniality, that
is, to the spatial theorization of modernity-coloniality and to
decolonizing action and thought. In this sense, they inform
geographies of decoloniality, which aim to understand socio-
physical relations arising from coloniality and decolonizing
actions.

Some may ask: how can geography decolonize concepts
and theories with such disparate inputs? Yet to ensure
geography transforms into a discipline appropriate for a
world 'where many worlds fit', this analytical plurality is
crucial. Indeed, acknowledging *plural* theoretical reference
points is entirely fitting, being consistent with decolonial
agendas to acknowledge and value multiple systems of
knowledge. The frameworks reflect differentiated geographies of coloniality-modernity, and diversely situated actions
to decolonize thinking. Moreover, pluralizing geographical

Table 2.1 Theoretical strands in geography's decoloniality
The table summarizes the four frameworks informing geography's decoloniality, the spatial theorization of modernity-coloniality, and decolonizing action and thought.

Theoretical framework	Key contribution to spatial theorization of coloniality and decolonizing thought and action	Influence in geography	Areas where the framework has particular traction and disciplinary influence
Postcolonialism	Colonial discourse and representations	From the 1990s, widespread influence across sub-disciplines from historical to urban and development geography	Nineteenth- and twentieth-century British (ex-)colonies in Sub-Saharan Africa and South Asia; urban and development thought and practice
Subaltern studies	Non-hegemonic spaces and subjects under colonialism; subaltern agency and voice	From the 1990s, influence in urban and historical geographies	South Asia and Latin America, historical and contemporary periods
Modernity-coloniality-decoloniality	Coeval modernity/coloniality; border knowledges; agency of subaltern	From 2005, influences geographical understandings of coloniality and epistemic decolonization	Latin America, Europe
Indigenous theories	Violences against Indigenous territories and peoples	From the 1980s, influences understandings of resurgence, refusal of settler sovereignty	North and South America, Australasia
Theories of settler colonialism	Settler logic of native elimination and creation of settler belonging; denial of pre-settler histories; territorial conflicts	From the 2000s, influences understandings of biopolitics, territory and sovereignty, and settler ideologies	North America, Australasia, Ireland, sub-Saharan Africa

Source: Author

conceptual-theoretical starting points adopts an ethical position by engaging with and learning alongside diverse scholars' contexts and concerns, and by decentring metropolitan theory. Geography's decolonial project to understand a complex world necessitates this conceptual-theoretical step of pluralizing its analyses and theorizations of modernity-coloniality, thereby rendering it better able to consider variegated colonial-modern realities across space and time (Radcliffe and Radhuber 2020).

Arguments for decolonizing currently have widespread acceptance in geography (de Leeuw and Hunt 2018: 5). Three decades after geography's 'cultural turn' (Gregory et al. 2009), the **'decolonial turn'** marks a moment when coloniality has become a central disciplinary concern. The decolonial turn demands broad cross-disciplinary action to decolonize geographical concepts, areas of interest, ethics, institutionalization, practices and interpersonal relations. More than simply an academic pursuit, decoloniality challenges the basic coordinates of modernity-coloniality in the pursuit of a humane world (Maldonado-Torres 2011; Daigle and Ramírez 2019). The next chapter discusses Black and Indigenous geographies, postcolonial-decolonial feminisms, and geographies of peace and violence whose critical analyses of coloniality-modernity directly challenge the discipline's whiteness and 'one-world world' Eurocentrism. These critical perspectives, as will be seen, offer key lessons to *all* geographers about understanding power, landscapes and geophysical processes, and about who we read and listen to, and to what purposes.

Further Reading and Resources

Readings

Bhambra, G. 2014. Postcolonial and decolonial dialogues. *Postcolonial Studies* 17(2): 115–21.
Blunt, A. and Wills, J. 2000. *Dissident Geographies*. Harlow, Pearson Educational.

Escobar, A. 2007. Worlds and knowledges otherwise. *Cultural Studies* 21(2–3): 179–210.

Jazeel, T. 2019. *Postcolonialism*. London, Routledge.

Sharp, J.P. 2009. *Geographies of Postcolonialism*. London, Sage.

Websites

Global Social Theory
https://globalsocialtheory.org
The Global Social Theory website has brief summaries on thinkers and concepts in decolonial theory, including modernity-coloniality, settler colonialism, border thinking and many more.

The Community Economies Institute
www.communityeconomies.org
CEI is a not-for-profit, member-based organization dedicated to furthering research, education and advocacy for economic practices that help us all to survive well together. The Institute works by cultivating and acting on new ways of thinking about economies and politics.

#IdleNoMore
www.idlenomore.ca/#IdleNoMore
IdleNoMore is an Indigenous-led Canadian social movement that calls 'on all people to join in a peaceful revolution, to honour Indigenous sovereignty, and to protect the land and water'. It has been at the forefront of resistance to petrochemical pipelines and gender violence against Indigenous women and girls.

–3–
Decolonizing Geographies

As Chapter 1 showed, the discipline of geography holds profoundly colonial views of the world. Imbued with the privileges of whiteness and 'one-world world' perspectives, geography faces pressing calls from students, academics and activists to acknowledge its colonial present and root out racism (**Chapter 1**). Decolonizing agendas are vital across both physical and human geography, although the challenges vary. Human geography has drawn on interdisciplinary postcolonial and decolonial social sciences and humanities, while physical geography is increasingly informed by critiques of colonial sciences (**Chapter 2**). The present chapter outlines a number of decolonizing paths to guide human and physical geography into delinking from colonial structures and knowledges.

Disengaging from coloniality is multifaceted and has no single endpoint; decoloniz*ing* takes geography in new directions without guarantees. All parts of the discipline need to address coloniality. Decolonizing offers an agenda that speaks to students, instructors and researchers alike because they are geographers who learn, teach and research in a colonial-modern context. With 'no clearly defined structure that neatly traces and binds decolonial geography' (Daigle and Ramírez 2019: 78), the open-ended processes of decolonizing geography connect to a core set of agendas, addressed throughout this chapter:

- To recognize the reasons for geography's coloniality (section I)
- To understand why decolonizing geography is necessary and possible, and applicable across the whole discipline (section II)
- To highlight geographical insights into modernity-coloniality and the basis for engagement with plural geographies (section III)
- To centre Black and Indigenous geographies, decolonial feminisms and critical geographies of violence (section IV)
- To reflect critically on the limitations of decolonizing approaches (section V)

The chapter ends with a summary of decolonizing geographies. Subsequent chapters develop the themes introduced here, specifically decolonizing geographical concepts (Chapter 4), decolonizing the curriculum and teaching (Chapter 5), and decolonial research practice (Chapter 6).

I Geography's Sanctioned Ignorance

Previous chapters showed how colonialism and its ongoing form of coloniality dismantled and suppressed diverse forms of knowing and being in the world. Conquest and occupation placed knowledges in a hierarchy, which elevated western science and understandings of the world above other longstanding systems of knowledge and experimental practice. This resulted in the destructive predominance of western forms of knowledge and permitted the Eurocentric one-world world to assume its universality (**Chapter 1**). Consequently, the west becomes less and less aware of or concerned about contributions from non-Euro-American thinking and practice. Geography is just as liable to ignore and sideline diverse insights as other disciplines. For instance, Arab, Hindu and Islamic navigation were well established by the mid-thirteenth century, long before European exploration began 200 years later, yet these knowledges are not incorporated into histories of geography (Sidaway 1997).

Whether in the sixteenth or twenty-first centuries, colonial-modern structures of power result in highly selective and skewed collections of knowledge and relations with the world. From vaccines to solar panels, a range of western scientific knowledges have proven their worth. Yet the world remains a one-world world in which alternative ways of thinking are dismissed as being based on merely descriptive or cultural beliefs. The postcolonial theorist Gayatri Spivak critiques this outcome, suggesting that globally powerful actors are in a position to *ignore* other knowledges because the latter appear to be irrelevant to the world. Moreover, this wilful ignorance – as the Black feminist Audre Lorde termed it – is actively *made* and *sanctioned* in colonial-modern domination and political economy (Spivak 1999). Other knowledges do not disappear, but are dismissed as marginal (**Chapter 2.II**). **Sanctioned ignorance** pervades university disciplines and teaching; universities end up teaching only sanctioned knowledges. Accordingly, 'academic practices and discourses … enable the continued exclusion of other than dominant Western epistemic and intellectual traditions' (Kuokkanen 2008: 60). In Anglophone and European contexts, physical and human geographies sanction their own ignorance about plural, vibrant ways of knowing, practising, and being in the world.

In some cases, western science does absorb other knowledges without necessarily recognizing them publicly, thereby erasing an understanding of non-Eurocentric knowledges as coherent modes of logic, practice and experimentation. For example, in crossing the Pacific Ocean, the British cartographer and explorer James Cook (1728–79) used western navigation charts, until he came across a distinctive system of navigation. The Tahitian navigator Tupaia (1725–70) adapted a Polynesian star-based system for Cook, creating a unique western-style, two-dimensional compass chart. On this and other occasions, Europeans selectively incorporated the sophisticated *fare-'di-ra'a-'upu* navigation method but

Figure 3.1 Micronesian tool for navigating by the stars
Source: Wikimedia Commons

denied its origins, thereby sanctioning ignorance of its
systematic components (Eckstein and Schwarz 2019; **Figure
3.1**). Under colonial-modernity, individuals such as Tupaia
– each with their own experience and finely honed practice
– have been drafted into colonial-era expeditions or PhD
projects without due recognition (Tuhiwai Smith 2006),
reproducing a western hierarchy between (Eurocentric) theory
and (Other) empirical knowledge, science and improvisation.
These hierarchies then shape how geographical information
is selected, categorized and fitted into what becomes known
as 'geography'. Over centuries, sanctioned ignorance about
marginalized realities has culminated in partial, misrepre-
sentative understandings of the world among publics and
professionals (Castleden et al. 2013; **Chapter 5.I**).

What insights might Polynesian star navigation bring to
geographers' efforts to understand ocean currents and the
distributions of plastic pollution? How might the practices
and transmission of star navigation inform an appreciation
of parallel and overlapping knowledge systems? Undoing
sanctioned geographies of ignorance is not straightforward,

as it challenges the ingrained tendency to overlook colonialism's and racism's pervasive consequences (**Chapter 1.II**). Decolonizing argues for the importance of unpicking the dominant narrative and questioning the dualistic nature/ culture framework that underpins physical/human geography divides. Critical decolonial geographies involve considering the interplay between geophysical and socio-economic dynamics. So decolonizing physical geography treats 'gender, racism, globalized markets [and] the repercussions of colonialism' as being 'as fundamental as the hydrological cycle, atmospheric circulation or plate tectonics' (Knitter et al. 2019: 459). Likewise, human geography must consider coloniality's geophysical transformations in relation to racial capitalism and systematic dehumanization.

These agendas are rooted in over five decades of geographical analysis of colonialism, racism and inequality. For example, the Association for Curriculum Development in Geography (ACDG), founded by the secondary school geography teacher Dawn Gill, has worked with schools and universities to challenge geography's racism and silence on apartheid (**Box 3.1**). The urgency of the human-induced climate change and mass extinctions called the Anthropocene leads physical geographers to study the political and socio-cultural drivers behind geomorphological change, and to recognize the diverse knowledges that can explain climate change (Slaymaker et al. 2020).

As discussed in Chapter 2, working towards decolonizing means looking critically at how coloniality values western **epistemology** (i.e. methods for knowing the world and transmitting knowledge) over non-scientific, non-western knowledge systems (Gregory et al. 2009: 206–8; Castree et al. 2013: 136). University geography overwhelmingly endorses and teaches western scientific epistemology in physical geography, and social science epistemology in human geography. As physical and human geographers acknowledge, overturning geographies of ignorance is never simply a question of increasing the *quantity* of information about the world. Rather the process involves dismantling hierarchies between different experiences of knowing and

Box 3.1 Anti-racism and school geographies in the UK

In the 1980s, a group of geographers campaigned against inequality and racism in the discipline and in society, prompted by the official rejection of a report on racism in the geography curriculum. Following uprisings in Brixton, London and Toxteth, Liverpool, the Schools Council was tasked with addressing racial issues, yet refused the report's finding that racism was widespread (Norcup 2015a). The group of geographers

Figure 3.2 Association for Curriculum Development in Geography 1983 conference, 'Racist society, geography curriculum'
Source: Norcup 2015b: 73–5 (Reprinted with permission from the curator and executors of the ACDG/CIGE archive)

then established the Association for Curriculum Development in Geography to challenge racism and the sanctioned ignorance of world affairs in classrooms. The Association's journal made demands for anti-racist and anti-sexist education, curriculum change and socio-environmental justice.[1] Their transformative vision of anti-racist geography curricula inspired teaching materials for all education levels on Apartheid South Africa, ecological crisis, war and peace, and gender and geography, tackling 'geographical knowledge-making's colonial past, its racial and western bias', and presenting 'arguments for postcolonial perspectives in geography' (cited in Norcup 2015a: 69; see also Norcup 2015b).

[1] A digital archive of the ACDG's journal is available at http://geographyworkshop.com/2018/03/29/archive-afterlives-digitising-cige.

between institutions. Decolonizing does not remove scientific approaches and techniques, but seeks to bring them into more horizontal relations with racially and colonially marginalized systems of knowledges, in order to create 'a better understanding of the power and knowledge structures of the system we have inhabited for the past 520 years' (Grosfoguel 2017: 147, 162).

Acknowledging multiple types and origins of geographical knowledges – 'a multiplicity of ... epistemological realities'

Epistemology

Epistemology concerns the ways we know about the world, what classifications we use, and the processes that confirm knowledge. Under coloniality, a dominant epistemology – such as that of European modern sciences – is based in institutions and worldviews that legitimate and reproduce certain knowledges at the expense of others.

(Mungwini 2013: 79) – is thus a key element of decolonizing. For Polynesians such as Tupaia, the world encompasses knowable relations between stars, ocean currents, waves, humans and other entities, which form a coherent knowledge system. In another example, non-Indigenous freshwater scientists at a Canadian research centre sat down together with Indigenous people to learn about each other's knowledge systems – a sometimes awkward process that nevertheless allowed the whole group to co-define the research undertaken (Bozhkov et al. 2020). The process of decolonizing geography occurs through these kind of transformations of power dynamics (de Sousa Santos 2014; **Chapter 1.VI**).

Due to global interactions and westernizing education, epistemologies do not exist in discrete, separate domains. Encounters – at times forced, often voluntary – continuously shake up ways of knowing and complicate the boundaries between standard ways of knowing the world and alternative ways. Tupaia's ocean chart, for instance, illustrates how formerly separate Pacific and European geographies interacted to generate a new way of thinking spatially. Tupaia's star-based system and his chart for Cook represent two facets of modernity-coloniality. Neither can be defined as 'tradition' or as irrelevant in the present because they exist(ed) alongside each other within colonial-modernity, although the Tahitian system was treated dismissively. This chapter argues that plural ways of knowing the world are lying in plain sight, and that geography can nurture a decolonial kaleidoscope of analysis by respectfully listening to and learning from path-breaking geographers.

II 'Alterable Geographies': Ways to Decolonize Geography

Having established that geography has been constructed through sanctioned ignorance, what persuades geographers to embark on decolonizing? Three issues have become central: first, geography's problematic status as an academic discipline; second, the need to work horizontally with plural forms of geographical knowledge; and third, demands for social justice. Together, these issues make the case for decolonizing across geography as a whole, as it is in urgent need of change.

Arguments for decolonizing rest on wide-ranging critiques of the discipline's colonial biases and presumptions. The fact that colonial impacts are durable and damaging was documented in the 1960s and 1970s by geographers focused on global anti-imperialism. They highlighted colonialism and dependent development, informed by geographical knowledge centres that actively sought to think differently to north Atlantic modalities. David Slater, for instance, worked at the University of Dar es Salaam in the 1970s, and throughout his career engaged with Brazilian, Spanish American and French geographers. At this time too, geographers identified colonial relations of territory and difference in 'fourth world' enclaves, where state and capital had expropriated Indigenous resources (Stea and Wisner 1984). Current debates around decolonizing geography highlight the challenges of changing colonialist thinking, undoing colonial epistemologies and facilitating plurality and horizontality (Noxolo 2017a; Esson 2018). The geographical status quo is maintained by the existing institutions of knowledge, largely (but not exclusively) north Atlantic, European and Australasian universities and institutes. Colonial-modern geography is grounded in powerful minds and places, whose epistemologies are supported by governments, elites and sectors of the public. By contrast, the voices, experiences and thinking of individuals and groups with plural, alternative understandings of the world and its geographies are either excluded entirely or hold less powerful positions in these institutions. Latinx, Black and Muslim geographers are minorities in English-speaking geography.

Decolonial arguments extend these concerns and ask how and why geography remains oblivious to and complicit in the colonial present. For Patricia Noxolo, we must answer two questions: 'first, how are geographers now inserting ourselves into ongoing dynamics of coloniality, and second, which particular aspects of the present moment are geography academics well placed to address?' (Noxolo 2017b: 318).

Addressing Noxolo's first question, geographers increasingly take note of the discipline's imbrications with all aspects of colonial power and of the inconvenient truth that geography actively sustains colonial relations (Holmes et al. 2014; Jazeel 2017). Responding to student and activist pressure, geography has to face up to its routine complicity with power, expressed in the one-world world and racialization. With respect to Noxolo's second question, geographers are well placed to pluralize geographies and analyse coloniality from human and physical perspectives (Mignolo 2000). For instance, as environmental geographies draw on geophysical and social science they can contribute decolonial insights regarding the kinds of voices and motivations behind discussions of the Anthropocene (Knitter et al. 2019; Dhillon 2020). To combat institutionalized exclusion, path-breaking geographers began to diversify geographical thinking beyond standard epistemologies. For example, the critical development geographer Janet G. Townsend worked closely with rural Mexican and Colombian women, documenting their lives from within a pluralized feminist framework (Townsend and collaborators 1994). Such engagements forge more horizontal, reciprocal relations in dialogues outside the university with publics and silenced groups. Re-centring the world on plural knowledges creates deeper engagements with non-metropolitan geography scholars, and facilitates learning from **Southern theory** to reformulate frameworks around non-western situations. This type of engagement requires of western geographers a thorough area-specific understanding and the ability to establish trust across differences (Jazeel 2014).

A third impetus towards decolonizing geography is the call for justice in light of colonial-modern inequalities and material inequities. In arguments that foreground geography's moral and ethical basis, decolonial debates express the urgency of solidarity and socio-environmental justice. These decolonial and ethical arguments energetically seek to minimize the boundaries between university geography and place-based activism, between scholars and diverse publics (Daigle and Ramírez 2019). For physical and human geographies, such commitments work in two directions: on the one hand, by supporting less powerful groups they validate demands for reparations, structural change and dignity (Fanon 2004; Esson et al. 2017); on the other, they redirect geography's purpose to 'enable liberation' (McKittrick 2019: 244).

In summary, arguments for decolonizing geography are based on a recognition of the institutionalized nature of colonial-modern knowledges, calls to work outside the university to create respectful and horizontal ways of learning with plural Other geographies, and action for social justice. In this sense, geography is not rigidly fixed in coloniality, but can be and is being altered and reoriented (McKittrick 2019). Decolonizing thus ultimately presents a hopeful agenda, holding out the possibility that by decolonizing *itself*, the discipline of geography can expand its anti-colonial actions and critiques, a point we turn to now.

III Turning the Decolonial into Geography

Having identified sanctioned ignorance and the rationales for decolonizing, diverse geographers have undertaken to develop a distinctively *geographical* lens to understand present-day colonial geophysical and social relations. While decolonizing 'work[s] to understand the world well beyond geographic disciplinary norms' (Holmes et al. 2014: 557), the skills and insights of physical and human geographers can map the operation of colonial-modernity. Coloniality-modernity produces 'multiple geographies, bodies and scars' (Pierre et al. 2020: 408), which in turn highlights the significance of the *spatial* dimensions of modernity-coloniality. Physical geography has begun to interpret 'stochastic and complex physical processes [that] shape the earth' (Knitter et al. 2019: 452) in relation to historic and contemporary political and social forces (**Chapter 4.III**). Decolonial human geography translates lessons from postcolonial, participatory and feminist geography into a critical lens to discern and analyse the socio-spatial dynamics of colonial-modernity across scales and spaces (Naylor et al. 2018). By making these moves, geography provides a discipline-specific analysis of coloniality's ongoing epistemic, embodied and material impacts.

Colonial representations of non-western people and places have been critically deconstructed by postcolonial geographers, who identify how visual content and discourse together uphold western self-identity and validate Eurocentric power. Inspired by Edward Said (**Chapter 2.I**), geographers demonstrate that imaginative geographies 'both legitimize and produce "worlds"' by and for geopolitical interests (Gregory et al. 2009: 369–71). Through texts and symbols, imaginative geographies dehumanize people and places 'elsewhere', making them intelligible through the western lenses of race, geopolitics and development (Gregory 2004). Paying close attention to geographical imaginations helps us to understand how hierarchies of difference (between the west and its Other) are produced relationally and become 'common sense'. For instance, North American projects to diversify food production and consumption draw upon (unacknowledged) white ideals, landscapes and histories originating in settler colonialism and Black plantation slavery and food provisioning (Guthman 2008; Ramírez 2015).

In comparison to other disciplines, geography brings a strongly contextualized and process-based understanding of dynamic, multiscalar, interconnected spatial processes in specific locations. To understand interlocking, mutually influential colonial-modern dynamics, interdisciplinary decolonial accounts track global processes over centuries, while geographers explain the multifaceted dimensions of spatial transformations across and within scales (Haesbaert 2021). For the Brazilian geographer Walter Cruz, decolonial geography analyses colonial-modernity 'on the ground', linking concrete multiscalar and specifically located processes (Cruz 2017). Studying climate change through a decolonial lens, for instance, means looking at political institutions and socio-cultural diversity, and how their interactions with complex geophysical processes shape the Earth. Human geography delves into how hegemonic geo-historic processes of coloniality shape the making of space and place (**Chapter**

4.I). For instance, Black people's experiences of housing, health and environment are rooted in enduring processes of enslavement, displacement, racial capitalism and the denial of Black agency (**section 3.IV**). Decolonizing geography withdraws from universalizing scientific approaches to refocus on situated spatial processes combining plural non-human entities, geophysical dynamics, life-worlds and multiple knowledge systems.

This means that what counts as decolonial geography is geographically differentiated, varying with location and situated geo-histories. Peoples' understandings of the world in relation to colonial-modernity and decolonizing are many and varied, as they arise from specific places and associated knowledges. In this sense, as two North American geographers suggest, 'The decolonial shapeshifts depending on the land you stand upon, including the differential decolonial desires layered into a place' (Daigle and Ramírez 2019: 78). Decolonial geographies counter modernity-coloniality by providing spatially disaggregated and contextualized accounts of dominant power, knowledge and subjectivities and spaces where coloniality is contested. Decolonial geographers often identify and affirm the place they speak from, as I do in this book's Preface. To challenge Eurocentric universalism, a geographer's institutional, social-political and epistemic positions are acknowledged. This recognizes that we each speak from a particular location, social milieu, institution and scholarly tradition. Two Canadian human geographers write: 'Michelle is Cree, a member of Constance Lake First Nation located in the Treaty 9 territory, and a new and uninvited guest on Musqueam territory. She is a new assistant professor in the UBC geography department ... Juanita is a white settler and uninvited guest on Musqueam territory. She is a tenured associate professor in the UBC geography department' (Daigle and Sundberg 2017: 341). Geographers' self-descriptions can change with dynamic shifts in social and personal positions, but they always avoid speaking from 'no-where'. Locating oneself in knowledge-power dynamics involves recognizing the awkward, compromised relations within coloniality and intersectional experiences.

What geography adds to decolonizing

This section has argued that geography's discipline-specific frameworks contribute in distinctive ways to decolonizing Eurocentric and colonial-modern processes and knowledge. Geography is a powerful tool both for analysing modernity-coloniality and for engaging respectfully with plural geographies and knowledges. Four geographical tools are important in this regard:

- Examining landscapes as geophysical and socio-cultural spaces which are powerfully shaped by colonial-modernity, and thus contested by different groups
- Understanding the enduring power of colonial-modern imaginative geographies
- Understanding modernity and coloniality as multiscalar and interconnected hegemonic processes that are spatially differentiated and geo-historic
- Acknowledging the insufficiency of exclusively western scientific approaches, and entering into dialogues with diverse forms and locations of knowledge

Decolonizing geography has begun to 'critique and reimagine decolonial theory through a geographic lens' (Naylor et al. 2018: 2). The next section builds on these insights by outlining the synergies between specific strands of critical geography and decolonial thinking.

IV Decolonial Kaleidoscope

How do we begin to think about geographical theory outside a Eurocentric worldview, consisting of mostly white geographers and philosophers, to not only learn but also honour other perspectives and worldviews that would include an engagement of both Black geographies and Indigenous knowledges? (Mahtani 2014: 365)

In answer to Minelle Mahtani's question, several strands of critical geography have shed significant light on directions for decolonizing geography. They share a scepticism about 'mainstream' geographical epistemologies, and deliberately work with and from standpoints, experiences and ways of thinking away from geography's one-world world. These endeavours hold lessons for the entire discipline as they reformulate understandings of relations between landscapes, peoples and processes. Together they offer multifaceted insights to what we can call a decolonizing kaleidoscope. The four strands of decolonizing geography discussed in this section are: Indigenous geographies and **Black geographies** (that is, work by and with Indigenous and Black individuals, including professional geographers, community knowledge-holders and intellectuals, and non-Indigenous/ Black allies); decolonial feminisms; and critical geographies of violence and peace. By critically analysing relations between bodies, ecologies, territories and geophysical processes, these approaches decolonize in profound ways our understanding of how the world works.

Indigenous and Black geographies are rightly viewed as foundational in geography's project of decolonizing, with lessons for both physical and human geographies. According to the British Caribbean geographer Patricia Noxolo, 'decolonization begins from the scholarship of black and indigenous peoples, and should be *led* by that scholarship' (Noxolo 2017b: 318, original emphasis). Indigenous and Black theory and foci emerged *despite* and *against* geography's western epistemologies, colonial science and whiteness. These geographies engage with geomorphology, fauna and flora, and diverse social groups from perspectives that are multiple and have only recently received due attention. For human geography, Indigenous and Black geographies are instructive about processes of dehumanization, non-dominant knowledges and plural alternatives in relating to and being in the world. For both physical and human geographers, Indigenous, Black and decolonial scholarship rethinks 'space and time [as] multiple and varied' (Naylor et al. 2018: 2), challenging understandings of causality, agency and process across scales.

All four decolonizing strands discussed here honour and uphold 'Indigenous [, Afro descendant and historically excluded] spatial knowledge and place-based practices on their own terms' (de Leeuw and Hunt 2018: 8). These spatial knowledges are irreducibly multiple and grounded in place, enacted and contested. Decolonial geographies can thus be defined as:

a diverse and interconnected landscape grounded in the particularities of each place, starting with the Indigenous lands/waters/peoples from which a geography emerges, and the ways these places are simultaneously sculpted by radical traditions of resistance and liberation embodied by Black, Latinx, Asian and other racialized communities. (Daigle and Ramírez 2019: 78)

As Daigle and Ramírez note, Black, Indigenous and subaltern geographies emerge in interconnection with particular places in the context of social justice demands, as well as in relation to landscape features and processes of interest to physical geographers. Recognizing these dimensions, biogeography has begun to delink from western epistemology to engage with Indigenous ways of seeing, categorizing and understanding plant biodiversity in intercultural exchanges (Bannister 2018).

Beyond Black and Indigenous geographies, decolonizing links to questions about the positions, experiences and thinking of social groups situated at the margins of the colonial matrix of power. Geographers stress the differentiated geographies made and nurtured by subaltern groups (Daigle and Ramírez 2019), and the intersecting colonial-modern relations of labour, gender, race-ethnicity, nationality and citizenship. Although Indigenous and Black subalterns are currently the most visible in decolonial geographies (de Leeuw and Hunt 2018: 3), colonial-modernity comes down harshly on many social groups, including migrants and refugees (Yuval-Davis et al. 2019), global care-workers, caste groups, racialized Asians, Latinxs, North Africans and minoritized religious groups (such as Muslims in historically

Christian societies) (Sidaway et al. 2014). Speaking from indigeneity, Blackness, intersectionality and violence challenges geography's default position, which is to consider the position of secure white, male, western citizens as the default norm. These critical approaches describe and analyse colonial-modern place, landscape and geophysical processes from the perspective of otherwise silenced and marginalized subjects and spaces. They thereby decentre our understanding of *who* does geography, *where* geographies are made, and *how* intellectual and political responses emerge. The four strands are discussed here in turn, although in practice they are interconnected.[1]

Indigenous geographies

Indigenous geographies – done by Indigenous and allied anti-colonial researchers – develop plural and decolonizing geographical topics. Anglophone western and settler epistemologies grant Indigenous knowledges peripheral status. Despite this, Indigenous geographers have initiated path-breaking discussions of anti-colonialism and settler colonialism in the discipline (Shaw et al. 2006; **Chapter 2.IV**). Indigenous geographies dispute powerful western narratives of Indigenous peoples as timeless and culture-bound, tied to local territory and nature, and instead give prominence to colonial-modern violences against territories, life-worlds and bodies (Johnson et al. 2007; Louis 2007; Coombes et al. 2012; Bryan and Wood 2015; Radcliffe 2017b). From this angle, Indigenous categories and **more-than-human** relations highlight the narrow conceptions that characterize most geography frameworks (Cameron et al. 2014; Todd 2016).

These starting points lead to important conversations. First, Indigenous geographies highlight the duress of coloniality in wealthy, white-majority nation-states such

---

[1] Despite their interconnected interests, these strands are in tension when the focus of Indigenous geographies on settler colonialism furthers anti-blackness. Black and Indigenous geographies are best considered in relation to racialization.

as Canada, the United States, Australia, Chile and New Zealand. Indigenous geographers make visible new spaces of capitalist expansion, the rapacious resource and labour exploitation across Indigenous territories, and the resistance against these processes (Howard-Wagner et al. 2018; Zaragocín 2019; Vela-Almeida et al. 2020; **Box 4.3**). In urban areas, Indigenous presence disrupts Eurocentric planning, reconfigures binary categories and becomes the ground for Indigenous place-making (e.g. Masuda et al. 2020). To disrupt colonial territorializations materially, Indigenous geographies examine and support plural relational human-territorial ethics and ways of knowing territory, ecologies and spirituality (Simpson 2011; Larsen and Johnson 2012). By these means, they contribute to Indigenous resurgence around grounded self-determination and expansive futures (Daigle 2016; **Chapter 2.IV**). Indigenous praxis dwells in land-territory and connects with living earth (L. Simpson 2014; **Chapter 4.III**), offering land-based teaching and learning experiences (Tuck et al. 2014; **Chapter 5.VI**).

Physical and human geographers need to recognize and honour diverse Indigenous worldviews that give equal parity to humans and non-humans, and shape behaviour and responsibilities accordingly. Physical geographers learning from Indigenous geographies broaden their understanding of reciprocal human-environment interactions. Indigenous land-based knowledges of bio-geographical and water-land dynamics can prompt physical geographers to rethink their western scientific explanations, not least as Indigenous geographies raise urgent questions about the destructive socio-environmental impacts of resource development (Bozhkov et al. 2020). Placing these epistemologies in dialogue, Indigenous scholars argue, generates 'two-eyed seeing' interpretations, which avoid singular narrow findings about geophysical change. For example, viewing soils as inert matter excludes thinking/knowing about soils as multi-species beings that care for/are cared for by human societies. In Brazil's Amazonia, the Canela group identifies ten soil types within an emergent bio-socio-cultural life-world (Miller 2019).

Indigenous geographers are wary of hegemonic interpretations of their realities, and frustrated when geography
re-centres settler voices, resists Indigenous theories and
permits anti-Indigenous violence (de Leeuw and Hunt 2018:
9). To overcome these challenges, Indigenous theories avoid
'culturalist' explanations that stress binaries, 'authenticity'
and tradition, and instead centre on structural processes that
vary over time and space. In this context, theorization from
diverse Indigenous perspectives seeks to recognize multiple
dimensions of social relations, even in one location. For
instance, the non-western forms of economy, authority and
knowledge of some Indigenous peoples are always already
enmeshed with colonial-modern power and universal notions
of difference (Briggs and Sharp 2004; Cameron et al. 2014;
Radcliffe 2020). Indigenous geographies are configured
integrally in relation to settler, Black and Indigenous groups
and non-native people of colour (Kobayashi and de Leeuw
2010; Pulido 2018). In summary, Indigenous geographies
document coloniality and Indigenous agency in order to
rethink social, spatial and geophysical processes from within
Indigenous places and geo-histories.

Black geographies

Geographers identifying as Black are underrepresented as staff
and as authors in north Atlantic and European universities
and the Antipodes, due to systematic exclusion originating
in colonial-modern racialization (Johnson 2018). Involving
both geographers identifying as Black and their anti-racist
allies, Black geographies think about how that exclusion
plays out in the real world and in the discipline. They draw
extensively on critical race theory and anti-colonial and
Black thought. In light of colonial-modernity's entrenched
anti-Blackness (Wynter 2003), Black geographies situate their
agendas for social justice across academia and public action.
These diverse geographies critically track the place-specific
configurations of racialization (e.g. how a neighbourhood
or job category becomes associated with one racialized
group) and anti-Blackness (McKittrick and Woods 2007;

Scott Lewis 2020; Wright 2020). By mapping ubiquitous yet locally configured patterns of anti-Blackness, Black geographies bring to decolonial geographies a powerful indictment of the violent ramifications of colonial-modern racism. These violences are expressed materially and in embodied effects, including premature death linked to urban segregation, environmental pollution, insecure jobs, classroom **microaggressions**, state-police violence and incarceration.

Black geographies pinpoint the origin of anti-Blackness in global colonial-modern structures and ideologies. 'Slavery and colonialism produced modernity as and with and through blackness' (McKittrick 2014: 17; 2019), constituting Blackness as a cluster of practices and meanings that foreshorten Black lives and disrupt or ignore Black spaces. Moving beyond 'universalist' history, this decolonial approach reframes historical geographies as entangled with enslavement and plantations (McKittrick 2011; Van Sant et al. 2020). Enslaved Blacks were uprooted from Africa and made placeless through systematic violence and the construction of new geographies of the plantation throughout the Americas and the Caribbean. The geographer Katherine McKittrick's concept of plantation futures foregrounds an enduring pattern of spatial control of Blackness in the Americas. The plantation, McKittrick suggests, 'fostered complex black and non-black geographies in the Americas and provided the blueprint for future sites of racial entanglement' (2011: 949) such as urban neighbourhoods and prisons (**Box 3.2**).

The deeply racialized logic of capitalism established land and labour systems situated initially, but not exclusively, in the Caribbean and Americas, and later in diverse diaspora spaces. Neoliberal capitalism reworks urban landscapes through by-laws and regulations in such a way as to re-embed racialized anti-Black exclusions in relation to mobility, interpersonal encounters and disputes over policing (Derickson 2017). The spatial distribution of environmental pollution disproportionately affects districts with Black and racialized populations, as they are less likely to command the resources or have the freedom to move to pollution-free places (Pulido

Box 3.2 Plantation futures: Katherine McKittrick

Black diasporic histories and geographies are difficult to track and cartographically map. Transatlantic slavery ... was predicated on various practices of spatialized violence that targeted black bodies and profited from erasing a black sense of place. Geographically, at the centre are the slave plantation and its attendant geographies. ... The plantation evidences an uneven colonial-racial economy that, while differently articulated across time and space, legalized black servitude while simultaneously sanctioning black placelessness and constraint. ... The conditions of bondage ... incited alternative mapping practices during and after transatlantic slavery, many of which were/are produced outside the official tenets of cartography: fugitive and maroon maps, family maps, music-maps were assembled alongside 'real' maps (those produced by black cartographers and explorers ...). ... The relational violences of modernity produce a condition of being black in the Americas that is predicated on struggle. ... The plantation notably stands at the centre of modernity. It fostered complex black and non-black geographies in the Americas and provided the blueprint for future sites of racial entanglement. [What] 'structures' a black sense of place are the knotted diasporic tenets of coloniality, dehumanisation, and resistance ... negotiating the power geometries of white supremacy. (McKittrick 2011: 948)

Having re-conceptualized Black geographies in the Americas, McKittrick then questions how to honour Black lives in the present without essentializing them or denying them agency. Focusing on the relational places of the home and the prison cell, McKittrick writes:

> Plantation futures [insist] that spaces of encounter, rather than transparent and completed spaces of racism and racist violence, hold in them useful anti-colonial practices and narratives. [Despite] the generalized traits of ... displacement, surveillance, and enforced slow death [in both plantations and prisons] ... prison life points to the everyday workings of incarceration as they are ... lived and experienced. ... Human relationality, rather than bifurcated systems of dispossession and possession, provides an important pathway into thinking through prison expansion. ... Put differently, we might re-imagine geographies of dispossession and racial violence not through the comfortable lenses of insides/outsides or us/them ... but as sites through which 'co-operative human efforts' can take place and have a place. (McKittrick 2011: 955; see also McKittrick 2013)

2000). Black geographies are always multiple, with diverse expressions of agency and liberation (Bledsoe and Wright 2019; Wright 2020). In rural Central America, for example, African-descent women negotiate for rights to land and livelihoods against systematic dehumanization (Mollett 2017).

The implications of Black geographies for the decolonizing of physical and environmental geographies are slowly becoming clear, although much remains to be done, including increasing physical geography's awareness of and action against anti-Blackness. The geographer Kathryn Yusoff (2018) argues that earth sciences are tightly linked to enslavement and dehumanization. In Southern Africa and elsewhere, Black labour serves to extract minerals in exploitative systems that cause high death rates (Esterhuysen et al. 2018). Using critical race theory, Black geographies affirm Black humanity, think about liberatory Black futures and seek restorative justice, thereby tracing decolonizing priorities for geography as a whole.

Decolonial feminisms and decolonizing embodiments

Decolonial feminisms in and outside geography also offer alternative lenses through which the discipline can examine colonial-modernity. They focus on intersectional relations, that is, on how racial, gendered, income, location and embodied differences become interlocked in the construction of hierarchies of power and social authority (Mollett and Faria 2018). Through the lens of gender and sexuality, decolonial feminists shed light on coloniality's basis in patriarchal domination and gendered norms. The Argentine feminist Maria Lugones suggests that coloniality's gender logic is deeply categorical, dichotomous and hierarchical, positioning all individuals in racial, gender, spatial and class classifications and power relations (Lugones 2007). The **coloniality of gender** involves not only the *imposition* of European gender hierarchies, it also actively dehumanizes through its techniques of control and its understandings of the non-metropolitan body and identity (Wynter 2003). Coloniality, says Lugones, sanctions the 'brutal access to people's bodies through unimaginable exploitation, violent sexual violation, control of reproduction, and systematic terror' (Lugones 2010: 744). Acts against human and non-human beings acceptable in one zone of coloniality-modernity are unacceptable in another, where intersectional relations are experienced differently (Grosfoguel 2019). Intersectional relations thereby connect the body to diverse geographies of colonial-modern power, knowledge and being (Mendoza 2015).

For Lugones, to decolonize is to critique 'racialized coloniality and capitalist heterosexualist gender oppression as a lived transformation of the social' (Lugones 2010: 746). Inspired by this goal, decolonial feminist geographies map the spatial arrangements that underpin hegemonic intersectional geographies (Mollett and Faria 2018: 566). In Latin America, the differential racialization of Indigenous and Black women confirms whiter men's entitlement to land and full citizenship (Mollett 2017). In Ecuador, Indigenous women dispute state development programmes that disregard their critiques of intersectional dispossession and dehumanization (Radcliffe 2015). Decolonial feminisms centre the experiences, knowledges and analytics of subordinated women in order to better understand the whole of society (Tomlinson 2018: 5; Hill Collins 2015). Indigenous feminisms depart from other feminisms and Indigenous theory in order to critique the colonial-modern operations of patriarchy in both Indigenous societies and white-centred feminisms. Indigenous feminisms struggle for (Indigenous) lands and (racialized female) bodies to be free of violent property-based colonial-modern control (**Chapter 4.II**). They also provide concepts and modes of working that enrich decolonizing climate science (Dhillon 2020).

A further dimension of decolonial critiques is their focus on the heteronormative dimensions of coloniality that shape power and space and structure inequalities. The coloniality of being results in the global dominance of western perspectives on queerness that marginalize and homogenize the realities of plural sexual positionalities. The combination of colonial legal bans on non-heterosexual acts and the promotion today of liberal gay rights in non-western countries exemplify these patterns. In response, queer and two-spirit Indigenous people in settler colonial areas seek to synergize anti-colonial and queer praxis and theorization (Hunt and Holmes 2015). Disablement is a further intersectional dimension of colonial power (Meekosha 2011). Colonialism impairs the embodied abilities of subjugated groups directly through conflict, environmental pollution, the global arms trade and damaging nutritional practices, while health policies reflect western 'universal' bodily assumptions and ignore coloniality's health impacts.

Decolonial feminisms and decolonizing thought about embodiments shed light on coloniality's expression through embodied intersectional relations of power. Starting from the intimate scale of the body, these perspectives decolonize western assumptions about the making of bodily difference and the nature of power, alerting human and physical geography to the ineluctably intersectional and colonial-modern making of experiences and perspectives.

Critical geographies of violence and peace

Critical geographies of violence and peace contribute to decolonial geographies' kaleidoscope by tracing how the spatial expression of structural and slow violences produces such unequal distributions of life and thriving. Geographical discussions of violence attend closely to the visible and less visible forms of brutality and intent that express structural and direct power. They identify the geopolitical, capitalist and colonial factors behind the interconnected violences found in mundane spaces (prison, property relations, resource extraction) and in everyday practices (Daigle and Ramírez 2019: 79). Drawing on the sociologist Johan Galtung's pioneering work, geographers examine how the structural violences of racism, patriarchy and classism play out over space and scale. Indigenous anti-colonial thinking sheds light on patterns of colonial violence in the present (Holmes et al. 2014). Colonial violence can be intimate and occur at the household scale, as in Canadian government interventions in First Nation families (de Leeuw 2013, 2017a). Additionally, slow forms of violence can take years or decades to materialize in harms to bodies and environmental damage, reflecting geographies of toxin release and accumulation, or housing insecurity and homelessness. Violence is shown to be bound up with the power relations of knowledge, as when officials use dominant knowledges and dismiss the cumulative experience and analysis of affected groups. By foregrounding the latter, decolonial geographers undo colonial frameworks and make alternative perspectives known (T. Davies 2019).

In seeking peaceful futures, geographers examine peace not as the opposite of the western liberal 'state hot war', but as a set of distributed, multiple and ongoing spatial processes (Courtheyn 2017; **Figure 3.3**). Peace can be seen in practices that support justice, dignity and care. Decolonizing therefore involves moving away from colonial-modern violences and towards relations that nurture people and places and allow them to thrive. These points have great relevance for physical geography, which historically has considered itself neutral and above conflict. However, acknowledging that conflict

Figure 3.3 Geographies of peace
In February 2017, Indigenous activists welcomed refugees and Muslim families at Los Angeles international airport after then-President Donald Trump's travel ban on arrivals from a list of predominantly Muslim countries. The Indigenous activists' intervention sought to create peace and hospitality. It also reflected decisions which asserted Indigenous power over the territory of Los Angeles (Daigle and Ramírez 2019; Walia 2012; Chapter 1.VI).
Source: Marc Nozell, via Wikimedia Commons: https://commons.wikimedia. org/wiki/File:No_ban_on_stolen_land_-WomensMarch_-WomensMarch 2018_-SenecaFalls_-NY_(24937353007).jpg

exists – including between western and other knowledge systems – is important. Decolonizing geomorphology, it is suggested, entails taking on board the fact that landscapes are contested (Slaymaker et al. 2020), as they reflect political, cultural and social processes of exclusion and privilege. Moreover, a decolonial lens brings into focus otherwise less visible non-'global' practices that contribute to reducing colonial-modern conflict. 'Decolonial struggles are struggles for peace, although they may not use that word' (Koopman 2019: 210). Physical and human geographers thus need the skills to work in contested intercultural spaces in constructive and peaceable ways.

In summary, these critical geographies of relegated experiences and marginalized epistemologies complicate narratives about living within and working against modernity-coloniality. Rather than reinforce white socio-spatial assumptions, these critical geographies refract and multiply the sources and kinds of geographical information available. The plural worlds they describe and analyse are neither all-encompassing nor commensurate with each other, and reflect enlivened and embodied places of knowing and unknowing (Pierre et al. 2020). Disrupting a western-centred world, together they generate a multi-coloured, compound and ever-changing *kaleidoscopic* understanding that works towards a decisive transformation by way of a hard-won critical knowledge. With this decolonial kaleidoscope, geography delinks from universalist compla-cency and gains insights into the double consciousness of modernity *and* coloniality.

V Ways Forward: Limitations of Decolonizing Approaches

As noted, decolonizing is a long-term process operating simultaneously *within* and *against* coloniality. Geography thus needs to continually reflect on the limitations and criti-cisms of where and how decolonizing is pursued. Criticisms to date focus on three points: the continued reliance on colonial-modern frameworks; the uneven decolonizing across physical and human geographies; and the limited leverage of academic decolonization in dismantling institutionalized coloniality.

Decolonial theory has received criticism for essential-izing and romanticizing non-western societies. The rush to see difference sometimes leads to an unquestioning reliance on colonial-modern classifications and analysis rather than on local experiences of and insights into complexity and dynamism. When insufficiently decolonizing, representations reproduce exclusionary biases, while overgeneralizations about subalterns sideline the dynamic indeterminacies of oppression and relative power. Here it is salutary to recall Homi Bhabha's (1994) argument that postcolonial identities are hybrid and ambivalent, as they mesh across colonial *and* non-western ways of being. Decolonization in these contexts means addressing that admixture and complexity alongside plural knowledge-holders (de Leeuw and Hunt 2018). In a Bolivian village for example, Guaraní inhabitants live within complex overlapping arrangements of subsistence and waged labour, in-flows of oil company money, activism for land rights and autonomy, and relations with spirit beings (Rivera Cusicanqui 2012; Anthias 2017).

Ultimately too, decolonizing has to encompass the full breadth of the discipline from human geographies through to diverse environmental and physical geographies. Whereas human and some environmental geographies are decolonizing, this is not mirrored to the same extent in physical geographical sciences (Lave 2015). Yet there is no intrinsic reason why decolonizing debates cannot invigorate physical geographies, as this chapter has shown. The biogeography and field ecology sub-disciplines illustrate how collaborations with diverse knowledge-holders can generate novel insights into living systems (Baker et al. 2019). Bryn Mawr's geology and colonialism reading list addresses relevant themes for geography (Bryn Mawr, undated). Topics include science and colonialism across world regions, doing science today in contexts of occupied land, Indigenous partnerships and land acknowledgement, as well as earth sciences' problems with whiteness.

Another criticism of decolonizing agendas, perhaps the most forceful, is that they speak loudly (especially in academia) but do little to disassemble exclusion and dispossession. At its mildest, decolonizing is seen as symptomatic of intergenerational squabbles over academic legacies (Barnett 2020). Certainly younger scholars often lead decolonial debates, yet, as this chapter has shown, anti-colonial, collaborative and Southern geographies have existed for decades, even if they were not always labelled decolonial. Moreover, many geographers advocate for decolonial analysis *and* action, bringing critical insights to environmental, anti-racism and equality activism. Stronger critiques view decolonizing as a slogan, offering an appearance of progressive politics without the hard and disruptive work to bring about material change in real worlds. Faced with centuries of exclusion, Black and Indigenous geographers question the willingness of and scope for white institutions and societies to effect deep change. Decolonizing in these terms means dismantling colonial structures to realize systematic and thorough transformations (Esson et al. 2017; Noxolo 2017a, 2017b). Indigenous movements excoriate university-based decolonizing for merely assuaging settler guilt. Eve Tuck and K.W. Yang (2012) argue that decolonization risks being a metaphor that masks settlers' self-justifying complacency, and that ultimately normalizes oppression and exclusion. In response to these critiques, geographers endeavour to be decolonial allies accountable to social movements in places where we work and live (Naylor et al. 2018: 4; on scholar allyship, see **Chapter 6.II**). In geography departments, allyship is crucial in reducing microaggressions and exclusion (Al-Saleh and Noterman 2020).

These are forceful and important critiques, prompting ongoing interventions into and reflections upon the direction and nature of decolonizing geography. In particular, they alert us to complex commitments and accountability in both teaching (**Chapter 5**) and research (**Chapter 6**).

VI Chapter Summary

The chapter has focused on what decolonizing does and should look like in geography. What are its strengths, what are the tools to hand, and what are its limitations? Addressing these issues revealed five dimensions of decolonizing geography, summarized in **Table 3.1**. These include the need to acknowledge that geography is situated in colonial-modernity's sanctioned ignorance. The history of modern-colonial knowledge and disciplines has sanctioned an ignorance of great swathes of human and earthly life, while validating Eurocentric claims to a universal understanding. Geography's sanctioned ignorance arises from its instrumental role in colonialism. Nevertheless, geography is amenable to change in decolonizing directions, and brings discipline-specific insights to modernity-coloniality's plural geographies. To honour contributions that critique and pluralize the discipline, the chapter introduced Indigenous, Black, decolonial feminist and violence-peace geographies. These critical approaches disrupt an 'earth-rise' view of a single world, and offer exciting bases for decolonizing understandings in geography, comprising a kaleidoscopic set of insights characterized by their plurality and patterned interconnections. The final concern of the chapter was to address head-on the limitations of decolonizing approaches. Non-western knowledges do not unilaterally replace western ones, yet assembling an 'open critical cosmopolitan pluriversalism' (Mbembe 2016: 37) presents numerous challenges, as **cosmopolitanism** is frequently Eurocentric.

Today geography has embarked on a 'critique and reimagin[ing of] decolonial theory through a geographic lens' (Naylor et al. 2018: 2). Although sceptical about *completing* decolonization, geographers remain hopeful of divesting the world and the discipline of their colonial legacies. Coloniality is a profoundly *spatial* phenomenon, as it arises through the articulation of places and peoples across diverse territories and ecologies at multiple scales. By contributing to decoloniality, geography plays an important role in wider decolonizing processes, because modernity and its underside, coloniality, are intrinsically spatial phenomena. Likewise, the processes that work towards decolonizing are 'diverse and located at multiple sites in multiple forms' (Sium et al. 2012: ii). Physical and human geographers are in unique positions to document and think through the real-world consequences of these decolonizing processes, which are as yet only partially realized.

Table 3.1 Decolonizing geographies: a summary

Why, how and what next?	Key points	Actions to decolonize
Reasons for decolonizing geography	The geography discipline currently condones and reproduces coloniality	Embark on multifaceted processes of critiquing colonial science (physical geography) and engaging with plural epistemologies (human geography)
	Need to open up to multiple, previously sidelined, experiences and thinking	Physical and human geographers learn from Black, Indigenous and decolonial feminist geographies, and from critical geographies of violence and peace
	Social justice: challenge structural racism and inequalities inside and outside the university	Sustained allyship and action
How to begin decolonizing geographies	Acknowledge the discipline's sanctioned ignorance	Engage in critical analysis of the colonial matrix of power, e.g. physical geography's basis in colonial science
	Rethink space, place and geography as plural and within coloniality-modernity	Approach coloniality-modernity as a pervasive, differentiated geographical process that shapes human, more-than-human and geophysical processes and outcomes
	Learn from Indigenous, Black and critical geographies, and decolonial feminisms	Open human and physical geographies to insights and knowledges from thinkers marginalized in the current colonial-modern world
Limitations of decolonizing frameworks	Essentialized or romantic view of non-western contexts	Listen to the relevant voices; acknowledge the heterogeneity and hybridity produced through power; give historic context
	Decolonizing is not yet mainstream across the entire discipline	Physical, environmental and human geographies are all implicated in colonial-modern relations of power and knowledge, so decolonizing is equally important across all parts of the discipline, although decolonizing debates and actions will vary
	Decolonizing risks being an academic buzzword, and sidelining action for structural change in institutions, society and mindsets	Build accountability and solidarity, and pursue material change

Source: Author

Further Reading and Resources

Readings

Baker, K., Eichhorn, M.P. and Griffiths, M. 2019. Decolonizing field ecology. *Biotropica* 51: 288–92.

Ferretti, F. 2019. History and philosophy of geography I: decolonizing the discipline, diversifying archives and historicising radicalism. *Progress in Human Geography* 44(6): 1161–71.

Radcliffe, S.A. and Radhuber, I. 2020. The political geographies of D/decolonization: Variegation and decolonial challenges of /in geography. *Political Geography* 78, https://doi.org/10.1016/j.polgeo.2019.102128.

Websites

Decolonizing conservation, a reading list
https://saracannon.ca/2020/06/27/decolonizing-conservation-a-reading-list
French and Spanish versions are available on the website too.

@DecoloniseSTEM
https://twitter.com/decolonisestem?lang=en
A community of scholars and science activists, exploring coloniality, privilege and power in and around science, technology, engineering and mathematics (STEM).

Geology and colonialism reading list
http://mineralogy.digital.brynmawr.edu/blog/geology-colonialism-reading-list
Lists sources on numerous themes including tectonics and earth science, as well as geography, biology and world regions.

Indigenous land acknowledgement
https://nativegov.org/a-guide-to-indigenous-land-acknowledgment
Indigenous peoples make the case for everyone to acknowledge their presence on Indigenous territory.

London Black histories map
http://www.blacklondonhistories.org.uk
Prepared by Rob Waters, historian at the University of Sussex and a member of the Raphael Samuel History Centre. 'This website is a digital mapping and blogging platform exploring histories of black London life, culture, and politics, covering the period from the 1958 white riots in Notting Hill to the 1981 Black People's Day of Action.'

–4–
Decolonizing Geographical Concepts

Building on Chapter 3, we discuss here the application of geography's decolonial kaleidoscope to the decolonizing of core geographical concepts. Over the past century, the use and careful analysis of concepts such as space, territory and nature have distinguished geography's contribution to wider knowledge and disciplinary theorization. Turning a decolonizing eye on classic concepts serves to reassess the presumptions and interpretations used by geographers to write about the world. Additionally, revisiting these foundational concepts helps us to understand the world in less exclusionary ways, and to develop plural frameworks. The Australian geographer Richie Howitt, drawing on his anti-colonial work with Aboriginal groups, argues forcefully for the remaking of geographical concepts to ensure the discipline forges a 'more solid and changeable engagement with complexity' (Howitt 2001a: 234). Absorbing the critiques of structural racism and of coloniality necessarily entails a thorough re-examination of core geographical concepts.

Focusing on English-language conventions and use, this chapter re-examines ten spatial concepts widely used in geography, developing decolonizing interpretations of each concept. Definitions are taken from the *Dictionary of Human Geography* (Gregory et al. 2009; now in its fifth edition), the *Oxford Dictionary of Human Geography* (Castree et al. 2013) and *A Feminist Glossary of Human Geography* (McDowell and Sharp 1999).[1] Providing authoritative introductions to geographical thinking on hundreds of topics, these dictionaries have been essential 'go-to' resources for generations of students and scholars, now complemented by *Key Words in Radical Geography: Antipode at 50* (Antipode Editorial Collective 2019).

The chapter aims to place decolonial interpretive frameworks within geography's mainstream vocabulary and epistemologies (Gregory et al. 2009: 206). Decolonizing geography also introduces new terms and definitions, reflecting inputs from plural geographies outside Anglophone scholarship. The eleven geographical concepts discussed are grouped thematically in the following sections: space, place, scale; society–space relations and territory; environment, landscape, the Earth and the Anthropocene; nature, the more-than-human and pluriverse.[2]

I Space, Place and Scale

Space is arguably one of the most significant terms in a geographer's vocabulary. Knowing about space – where things are distributed across the Earth's surface – has been geographers' 'unique selling point' and distinctive contribution since the early modern-colonial period. Geography has long claimed expertise in knowing what space is, from explorers and cartographers mapping locations on the Earth's surface with European cartographic conventions (Gregory et al. 2009) through to GIS applications today. Drawing on decolonizing approaches, we can begin to understand how Eurocentric ways of conceiving space came to dominate.

---

[1] These geography dictionaries broadly encompass the remit and scope of English-language geography; contributors are largely based in Anglophone and settler colonial countries.

[2] This selection omits concepts such as region, frontier and the urban due to lack of space (postcolonial-decolonial perspectives on these themes are addressed in Parnell and Oldfield 2014; Robinson and Roy 2016).

Two Enlightenment-era thinkers were particularly significant in shaping Eurocentric notions of space and the comparisons between spaces (and people) across the world, namely Kant and Hegel. The German philosopher Immanuel Kant (1724–1804) contributed in numerous ways to geographical conceptions of space and human agency (Gregory et al. 2009). Significantly for decolonizing approaches, his view of space – with externally defined coordinates resulting in 'phenomena beside each other in space' – as a human universal underpins colonial viewpoints. Coloniality treats space as a 'container' and arena, which in turn informed colonialism and empire (Blomley 2003; Law 2015). Furthermore, Kant associated the tropics with Blackness, imbuing the concept of space with racialized vocabularies of power and difference (Kobayashi 2002).

Informed by such notions of abstract space, colonial law used doctrines of discovery to designate large swathes of the Earth's surface as *terra nullius* (Latin for 'land of nobody', that is, without legitimate human occupation), thereby opening it up for the clearance of existing inhabitants and occupation by European settlers.[3] A second key influence on modern European conceptions of space and the relative standing of different societies was G.W.F. Hegel (1770–1831), another German philosopher. Numerous postcolonial and Black scholars have critiqued Hegel's notion that Africa did not have philosophy and that Indigenous displacement was justified. Yet it is his abstract formulation of Eurocentric human self-determination and its epistemic power to determine others' lives that deservedly generates geographers' sharpest postcolonial-decolonial criticisms. Hegel's ideas about the unfolding of human possibility across the globe have had enduring influence and colonial-modern consequences. For Hegel, human potential was uniquely European, a self-legitimizing quality that justified interventions in non-European spaces and societies (Power 1998).

---

[3] The eighteenth-century legal doctrine of *terra nullius* allowed colonial settlers to legally claim land if it was uninhabited or used unproductively – both criteria that drew on Eurocentric understandings (Blomley 2003; Wolfe 2006).

Today, Hegel's ideas underlie development theory's notions of progress and wider claims about 'the end of history'. To decolonize Hegel, the Black geographer Jovan Scott Lewis (2018) calls for multicultural, multilingual 'circular, spiral relationships' and a diaspora epistemology delinked from universal space.

By the 1960s, Euro-American critical geographers were beginning to rethink space as something produced through interlocking economic, political and social processes, resulting from capitalism's production of a 'space economy' (Harvey 1973; Lefebvre 1991). Geographers increasingly acknowledged and tracked the differentiated and relational ways that social groups (defined by class, gender and 'race') accessed spatial resources (property, land, etc.), traversed space (McDowell and Sharp 1999) and were divided by socio-spatial patterns of inclusion and exclusion. However, space was theorized primarily in relation to European and North American conversations, drawing on Marxist and poststructuralist theories that left Euro-America unmarked and the global South pathologized, while 'race' was analysed in terms of non-white groups, such that geography's theorization of space erased whiteness (Bonnett 1997; McGuinness 2000). Despite this era's emphasis on power, differentiation and contingency, space was rarely considered in relation to coloniality-modernity or to the privilege and normalization of whiteness (cf. Slater 2004).

At the same time, geographers in Brazil were rethinking space and time in decolonizing ways. Milton Santos (1926–2001) led debates that highlighted the irreducibly place-specific configurations of space constituted within uneven and hierarchical global relations (Santos 1977, 1995). Since then Brazilian geography has developed a highly pluralized understanding of space in relation to modernity-coloniality. Cruz (2017) views space-time in terms of diverse, multiple places, territories and cultures, each of them with distinctive trajectories and plural ontologies. In other words, decolonizing space treats 'space' as being as intrinsically plural as the knowledges and subjectivities that relationally constitute it (Daigle and Ramírez 2019: 79). In these interpretations, space is not universal and abstract, but produced through grounded **epistemes** and practices. To take one example, Australian Aboriginal groups understand 'space' (not their term) as co-created by interacting forces of topographic features, plants, animals, ritual sites and ancestral beings, each integral to the emergent 'being' of space (Bawaka Country et al. 2016a, 2016b). Another strand in thinking space decolonially switches to conceiving space as producing inclusion rather than entrenching inequalities. In this vein, the Black geographer Katherine McKittrick calls for geography to examine 'how, where, why, and when the production of space enables liberation' (McKittrick 2019: 244). For example, Latin American Indigenous communities produce spaces where they practise autonomous forms of decision-making and justice, liberating them from systematic exclusion and discrimination.

Ontology

Ontology concerns the nature of 'being' and that which can be said to exist in the world. Decolonial approaches highlight the location and socio-historical specificities of ontologies. Anti-colonial critiques of western ontologies identify the processes and institutions by which non-western ontologies are devalued and undermined.

Space in its relation with time has been a further arena for decolonizing geography. Colonial-modern power was expressed in the means used to coordinate space-time around Eurocentric interests. After the establishment of Greenwich Mean Time (noon at zero degrees longitude) in 1675 in London, colonies and empires ran to western time-pieces. With decoloniality, geographers are pluralizing their understanding of space-time, and considering the social, political and cultural ramifications. Indigenous groups are often accused of living in the past, or are not expected to survive into the future. One counter to these arguments comes from Pacific Islanders, for whom Hawaiian 'histories, memories and stories [are] in our skin' and contribute to an inclusive and sustainable future time (Kamaoli-Kuwada 2015). Rather than moving into a Eurocentric modern future, Indigenous concepts of space-time reflect place-specific configurations of coloniality and particular political visions. In Aotearoa-New Zealand, Māori space-time – *Wā* – comprises a multidimensional matrix in intimate connection with ancestors across (western) past, present and future. Māori young people occupy and access this supportive space-time as they make their way across segregated (white/non-white) cities using particular practices to counter settler colonial relations (Kidman et al. 2021). In North America, Indigenous nations counter the colonial matrix of space-time (capitalist work time, short-term decision-making) by invoking kinship relations with land, 'seven-generation' time-frames and non-human temporalities, for example in action to oppose pipeline development across territories (Awâsis 2020). In the Andes, the concept of Pachakuti refers to multiple, superimposed, colonially disrupted notions of space-time. For the Andean observer, the future lies 'behind' (as it remains invisible) and conquest is 'in front'; these conceptions inform social movements to decolonize governance. These examples highlight the need to take place-specific understandings of space-time into account in decolonizing geography.

Like space, the notion of *place* has played a unique role in geographical thinking. While in a minimal sense the term refers to a fixed point or locale of any size or configuration (Castree et al. 2013; Gregory et al. 2009: 539), it has acquired further disciplinary meanings which reflect Eurocentric thinking. Place is associated with human transformations of the Earth's surface, creating differentiated locales each with social-cultural meanings and practices. Differentiated relationships with place – according to interconnected hierarchies of gender, 'race', sexuality, age, etc. – have been extensively documented (McDowell and Sharp 1999: 201). In Marxist and economic geography perspectives, place reflects a 'particular moment within the production of space', as economic processes must be grounded in locations as they compete in a global economy (Gregory et al. 2009: 540).

Although globalization interconnects places, they remain unique due to their configurations of identities, institutions, flows and everyday forms of life. The British geographer Doreen Massey's 'progressive sense of place' shows how place retains its significance, while constantly adapting in an age of uneven development and global interconnections.[4]

Decolonizing geography, however, prompts interrogation of the ability of white epistemic power to make itself invisible even in supposedly critical accounts of place (McGuinness 2000). Indigenous, Black and border knowledges question 'white-blind' approaches, focusing instead on how place is made by overlapping, located configurations of colonial-modern ideologies, structural exclusions and white/Eurocentric place-making. Key to their critiques are arguments and evidence that place is made in and through colonialism, epistemic violence and dehumanization. Imperial economic interests forcibly moved subordinated groups from their existing places to places made by and for modernity-coloniality. Black enslavement depopulated meaningful African places and enrolled millions into Caribbean, North and South American plantations. Frantz Fanon vividly described how colonial anti-Blackness 'disconnect[s] the black body from its place in the world' (Kobayashi and de Leeuw 2010: 131). South Asian indentured labour was moved half way across the world, Chinese labourers were transported thousands of miles to build infrastructure, and Andeans were transplanted into Spanish colonial towns for surveillance. Incoming Europeans were authorized as 'place-makers', using surveys and grids and introducing plants and new social relations (Blomley 2003). Settler colonial place-making occurred within legal, narrative and social relations resulting in Indigenous dispossession (**Chapter 2.V**). In areas cleared of Indigenous inhabitants, western forms of settled agriculture were implanted (**Box 4.1**). The largely unspoken

---

[4] Doreen Massey's work is widely translated and circulated; her overseas students and global South geographers in turn develop geographical Southern theory (Connell 2007).

Box 4.1 Indigenous place-making under settler colonialism: Oklahoma

In July 2020, the US Supreme Court ruled 5–4 that a vast area in eastern Oklahoma state is an Indigenous reservation, including much of the state's second city (**Figure 4.1**). The case considered whether the Muscogee (Creek) Nation lands legally remained a reservation after Oklahoma state joined the USA in 1907. It illustrates how settler colonialism resounds into the present by constructing landscapes, political geographies and geographical imaginations. The Muscogee (*Este Mvskokvike*) Nation is the fourth largest US tribe, with some 86,000 enrolled members, most of them in Oklahoma. They are a federally recognized Nation that has governed housing, health and social programmes, and tribal police since 1970. The court judged that the Muscogee Nation's reservation (allotted in the nineteenth century) 'has not been disestablished'. The ruling upheld the original terms of that allocation specifying that the lands would be secure forever. Oklahoma has the second highest number and percentage (8.2%) of native Americans of US states according to 2002 figures (also 7.3% Black-African American, 1.7% Asian, 8.9% Hispanic-Latino) (Guardian 2020).

The court ruling came in the wake of forced removal, slow and acute violence, and intergenerational trauma for the Muscogee and other Nations who were expelled to lands west of the Mississippi. The Trail of Tears for Muscogee, Seminole, Chickasaw, Choctaw and Cherokee Nations and their Black slaves resulted from forced evictions and long treks, causing many lives to be lost. An evolving white-settler farmer policy and racialized capitalism, including the government's Louisiana Purchase and a succession of anti-Indian policies, culminated in the uprooting of over thirty diverse Nations from their places in the southeast and, sixty years later, their relocation in 'Indian Territory' in what is now Oklahoma.

Figure 4.1 Settler colonial geographies and Indigenous places in Oklahoma
Source: Department of Geography, University of Cambridge

The nineteenth-century allocation of a reservation for the Muscogee, however, was far from secure, and was continuously restructured in light of settler interests, with systematic moves to disband native treaties and repopulate reservations with European-descent settlers. From 1871 no further treaties were drawn up, while legal and federally sanctioned actions abrogated existing treaties (Wolfe 2006). The 1887 Dawes Allotment Act required communal tribal land to be subdivided into smallholdings for non-native settlers, with 'excess lands' reverting to the federal government. The 1898 Curtis Act dismantled tribal governments. Indigenous-held lands declined from 52 million acres to 15 million acres between 1881 and 1931. Forced assimilation into white society 'attacked every aspect of native American life – religion, speech, political freedoms, economic liberty, and cultural diversity' (Wolfe 2006: 400). In this context, native leaders and scholars speak out against racist stereotyping and take legal action to enforce treaty rights. Complex relations between Indigenous, settler and Black populations continue: in 2007, the Oklahoma Cherokee voted overwhelmingly to eject descendants of former Black slaves from tribal citizenship.

power of whiteness to shape places has accordingly come under closer scrutiny.

Despite coloniality's places, however, alternative place-making operates at a remove, where practices and identities are not drawn exclusively from colonial-modern power. Dominion over space and territory achieved through political technologies and violence have not quenched the agency and cultural-political distinctiveness of subordinated groups. Beyond the interest or reach of colonial-modern authorities, identities and markers of place distinctiveness have emerged, rendering 'place ... the site of the subaltern par excellence' (Escobar 2007: 199). Dominated groups must 'make place' in insecure but meaning-laden locations even as colonial-modern processes seek to displace them. Appearing from modernity's underside, these places express hybrid forms of place and meaning. As Indigenous and Black geographers document, less powerful racialized groups forge meaningful places and identities, such as Black senses of place that have been complexly constituted in tensioned sites of slave plantations and the prison system (McKittrick 2011; **Chapter 3.IV**).

Indigenous and non-Indigenous geographers add to the decolonial understanding of place by drawing attention to the human and more-than-human nurturing of place by means of human interrelations with the Earth's surface and features (Murton 2012; Barker and Pickerill 2019). In a physical geography example, soils and soil-plant combinations are an integral component of Australian city life that is often ignored. By contrast, Aboriginal relations with more-than-human Country (encompassing land and tangible and non-tangible elements) highlight the closely intertwined relations between human and more-than-human beings that bring soils, plants and place elements into being (Darug Ngurra et al. 2019). As Indigenous peoples face ongoing coloniality, their places are situated in the cross-hairs of modernity-coloniality and resistance, reflecting 'the myriad ways Indigenous peoples think about and live self-determination outside and/or alongside of formal state and intergovernmental structures' (Daigle 2016: 261). In these interstitial places, decolonizing accounts reassess Eurocentric understandings of place as a 'canvas' for human agency. Instead they raise questions about the differential agency of humans and more-than-human beings (Larsen and Johnson 2016; Clement 2017). Places, in other words, do not reside within the singular universe of a Eurocentric one-world world, but represent a 'fractiverse' (Law 2015) or a **'pluriverse'** (**Box 4.5**).

Throughout its disciplinary history, geography has placed considerable emphasis on *scale*, first in relation to cartographic scale or resolution in maps, and then in methodological distinctions between levels, from the body (micro-scale) up through the scales of neighbourhood, district, city, region, nation and globe. Recently, geographers have begun to think about scale not in terms of predetermined levels but in terms of the *outcomes* of spatial processes that are uneven, dynamic and inherently power-laden (Gregory et al. 2009: 664–5). Viewing scale in this way resonates with critical race and decolonial geographers as it can pinpoint modern-colonial processes and unequal power, knowledge and social relations. Coloniality focuses its discourses on one scale in ways that naturalize a narrow interpretation of wider structural issues. For example, when dominant representations of 'race' focus on the urban neighbourhood scale they treat race merely as a local issue, whereas racism operates dialectically at multiple scales encompassing the 'individual, the group, the institution, society, the global' (Pulido 2000: 15). Yet relations across these multiple scales can also contribute to decolonizing action. For example, in June 2020, Black Lives Matter anti-racist action erupted at multiple scales, on the streets of Minneapolis where George Floyd was murdered, across the USA, and at the global scale. In different circumstances, Andean and Amazonian Indigenous networks 'jump scale' to counter global development and political economy, thereby overturning imaginaries of agency-less 'local' residents. Decolonial action thus develops a 'scalar consciousness', as with the Zapatistas in southern Mexico, who deliberately organize alternative political governance in local 'caracoles' and regional assemblies (Harvey 2016).

In summary, decolonizing approaches encompass a plural sense of the world which changes the analytical meanings of the core geographical concepts of space, place and scale. These concepts then come to be understood as inextricably shaped by colonial-modernity. Decolonizing scale extends critical deconstructions of local–global binaries and leads towards a multiscalar analysis of subaltern dynamics of producing space, place and scale. An openness to alterity and multiplicity situates these concepts outside a universal framework, and centres differences of standpoint and conceptualization.

II Society and Space

From the 1980s onwards, **Anglophone geography** began to understand that society and space are mutually constituted rather than separate self-contained units on a surface. In this context, the *human subject* came under increasing scrutiny in an effort to discern how and why social differences were expressed in spatially differentiated ways (along class, gendered and racialized lines), and how relations of power co-constituted social and spatial relations (Gregory et al. 2009). Decolonizing the human subject alerts us to the epistemic and psychic violence that works to differentiate human subjects.

Decolonial approaches deepen the postcolonial critiques of Eurocentric conceptions of the human subject, and resituate analysis in relation to colonial-modern power and the experiences of border subjects. As described in Chapter 2, postcolonial analysis demonstrated the enduring significance of colonial understandings of human socio-spatial relations, which perpetuate processes of Othering. Taking this forward, decolonial thinkers point to the psychic violence that exists in modernity-coloniality, and which gives rise to qualitatively different ways of being human (Fanon 2004; Maldonado-Torres 2007). Thus, for Fanon, there is no single, universal human subject, because an anti-Black world results in a radically different experience of being for Black subjects in comparison to western white subjects (Wynter 2003). Moreover, these relations of violence are expressed spatially, as colonial-modern racialization produces 'zones of non-being [which have] a spatial manifestation' (Grosfoguel 2017: 154).

Distinctions between subjects relate to colonial-modern hierarchies of humanity, as well as relational distinctions separating humans from non-human animals, wilderness and nature. The zones of non-being for racialized women in the Americas, for instance, are tied to specific places, whether Canadian residential schools for Indigenous peoples, or extractive mining zones in the Amazon (**Box 4.2**). Colonial-imperial hierarchies of human being and human 'non-being' are also forged in relation to wildness and animal qualities (Hovorka 2017). Colonial encounters with Aboriginals in eighteenth-century Australia disturbed European distinctions between humans and nature and unsettled assertions that humans were 'nature-transcending' (Anderson 2008: 165). With regard to gender, qualities considered to be less than human have been used to justify direct and epistemic violence against racialized women, undermining the security of their bodies and landed property (**Chapter 3.IV**). In Latin America, Indigenous women were positioned as 'rape-able' while enslaved Black subjects were considered closer to non-human animals than to humanity (Mollett 2017). In summary, decolonial discussions point to how coloniality-modernity severs connections between humans, by denying full humanity and flourishing to subaltern subjects-in-space.

Key to Anglophone geography's accounts of society–space relations is *territory*. Western political thought, notes the geographer Stuart Elden, defines modern territory as 'a bounded space under the control of a group of people, usually a state' (2013: 322). Emerging from society–space relations, territory comprises 'a unit of contiguous space that is used, organised and managed by a social group, individual person or institution to restrict and control access to people and places' (Gregory et al. 2009: 746). Anglophone geographical accounts of territory are primarily based on a reading of *western* history, theory and concerns (Halvorsen 2018). Decolonizing the concept involves unpacking how colonial-modern power and knowledge shaped and still

Box 4.2 Murdered and missing Indigenous, Black and racialized women and girls in the Americas

Across the Americas, racialized women are subjected to gender-specific and deliberately degrading violence that links to coloniality's geographies. In Canada, some 4,233 Indigenous and First Nations women were murdered or went missing between 1980 and 2016. Although making up less than 5 per cent of the population, Indigenous women and girls account for 16 per cent of female homicide victims. After years of activism, national Indigenous women's organizations successfully campaigned for a national inquiry. The inquiry worked within settler colonial spaces and notions of justice, evidence and representation, to the disappointment and anger of Indigenous women's families and movements (Lucchesi 2019).

In Latin America, gender-based violence against and masculinist attitudes towards racialized women date back to the colonial era, materially reproduced in states, armed actors and domestic spaces today. Structural and symbolic violence reproduces a continuum of violence against women, reflected in sexual intimidation and rape, especially against Indigenous, Afro-descendent and mestiza women. Dominant and subordinated men adopt aggressively masculine practices, which pervade state security and para-military organizations. Femicide – the killing of females by men because they are female – is now socially condoned and legally permitted in many Latin American countries, with state agents failing to take action. In Guatemala for instance, Maya Indigenous women were killed at twice the rate of non-racialized Ladina women during the civil war (1960–96), and nine in ten rape victims are thought to be Maya women, whose lives become expendable and their bodies spectacularly defiled (Carey and Torres 2010).

shape territory, and acknowledging alternative ways of making and conceiving it. The modern European nation-state, which emerged in seventeenth-century Westphalia in northwest Germany, was based on sovereign control over territory. As such, it inaugurated an exclusionary notion of territory which, according to decolonial geographers, made it susceptible to intolerance to those outside the territory and facilitated racialized colonization and settlement (Porto-Gonçalves 2017). By contrast, decolonial discussions of territory use a broader conception of power, not limited to the state's coercive power, but taking multiple forms (Haesbaert 2011, 2021).

Colonialism historically displaced and continues to displace populations and prior modes of inhabiting territory. A decolonizing perspective alerts us to meanings and spatial practices other-wise, by reorienting analysis towards the plural and contested forms territory has for diverse groups. Rather than fixing on a singular territory, the Brazilian geographical concept of multi-territorialities refers to politically, epistemologically and socially diverse projects of territory-making, some expressing state or colonizing notions, others with hybridized and heterogeneous territorial relations (Haesbaert 2013). Decolonial feminist geographers argue that the concept of body-territory (*cuerpo-territorio*) resituates agency and colonial-modern harms in a relational dynamic (**Box 4.3**). Decolonial approaches view territory as relational, processual and plural.

III Environment and Landscape

In Anglophone geography, *landscape* has two primary – slightly overlapping – meanings. First, landscape can refer to pictural representations of nature; second, geographers view landscape as a material hybrid of culture and nature that reflects the selection, disposition and meanings of elements through power relations (Gregory et al. 2009). For example, suburban New England planning laws require manicured lawns and white fences, which in turn requires workers

Box 4.3 Body-territory/*Cuerpo-territorio*

Indigenous women and decolonial feminist geographers across Latin America lead discussions and activism around what they term 'body-territory' (*cuerpo-territorio* in Spanish). From Ecuador's Amazonian oil fields and Indigenous territories, diverse women marched to the capital Quito in 2018 demanding protection of lives and lands. As elsewhere, this multi-ethnic group are experiencing a rapid expansion of mining, infrastructure development and hydrocarbon extraction, which cause spikes in male employment, masculine use of public spaces, and patriarchal behaviour. In these contexts, violence and environmental degradation processes systematically reduce women's and Indigenous groups' decision-making power.

Indigenous women understand their bodies and their territories to be integrally interconnected, and they seek to defend the autonomy of body-territory: neither women nor land are territories to be conquered (Zaragocín 2019). Body-territory refers to bodily and territorial entities brought together materially and embodied through oppression and in resistance. Body-territory comprises webs of diverse living socio-natural entities that resist destruction and opt for life. Latin American feminist geographers argue that the concept decolonizes Anglophone geography by rethinking the body and territory in conjunction. Body-territory 'supersedes the individual and resituates it within a communal subject agency ... [it] prioritizes the body as the unit of analysis of spatial dynamics' (Zaragocín and Caretta 2020: 5).

In North America, coalitions of Indigenous women and First Nations young people mobilize against environmental violence. Mining areas across the United States and Canada experience rates of gender violence rising in line with extraction, in turn exacerbating intersectional violence against Indigenous women and young people. The Bakken oil boom in a largely Indigenous area of North Dakota triggered an 18 per cent rise in crimes including sexual assault, domestic disputes and disappearances. Levels of violence against Indigenous women were *already* 2.5 times higher than for any other group of women, while the majority of assailants are non-native.[1]

In North and South America, the decolonial concept of body-territory focuses attention on expressions of coloniality at multiple scales in socio-spatial processes (capitalist extraction, male migration, racialized female bodies) and colonially inflected sites (Indigenous reservations and extraction sites). Body-territory decolonizes Anglophone geography as it expresses the knowledge and experiences of those at the margins of colonial-modern processes.

[1] See landbodydefense.org.

to garden and maintain appearances that reflect European aesthetics (Duncan and Duncan 2004). Marxist geography highlights the largely invisible hard labour required for digging, clearing and trimming the landscape. More recently, geographical approaches have drawn attention to movement through a landscape in terms of bodily and emotional dimensions (Wylie 2005). Decolonizing critiques reframe these concerns to understand the colonial-modern formations of landscape. Critical race and decolonial scholars foreground the power of whiteness to constitute landscapes where people of colour are rendered 'out of place'. For instance, New York's Central Park was created in part by clearing the homes of African-Americans and Irish to construct a 'natural, sanctifying, wholesome, and White [landscape], counterposing it against a city construed as artificial, profane, insalubrious, and coloured' (Byrne and Wolch 2009: 747). The ongoing racialized coloniality of landscapes means that non-white people become hyper-visible in white spaces, as in British national parks and wilderness.[5]

Nevertheless, the material and symbolic dimensions of landscape also express subaltern agency and plural social-epistemic positionalities. Landscapes are meaningful for oppressed groups in ways that do not correspond to dominant understandings. Histories of colonialism are inscribed into landscapes, yet ongoing coloniality is resisted through practices of working with the land to construct decolonial meanings. For some Indigenous groups, land is synonymous with place, comprising a meaningful home that encompasses the physical landscape as well as water, sky, plants, humans, spirits and other components (Barker and Pickerill 2019). In Aotearoa-New Zealand, Māori people name features of landscape that are simultaneously ancestors, thereby evoking geographies that exceed standard geographical interpretations (Tuck et al. 2014). North American Anishinaabe think 'we are made from the land; our flesh is literally an extension of soil' (Vanessa Watts quoted in Davis and Todd 2017: 769). Taking into account plural landscapes destabilizes Anglophone conceptions of a passive and non-human natural landscape. For Chatino Indigenous people in southern Mexico for instance, landscape encompasses the world and human/ more-than-human life and thus problematizes colonial-modern binaries (Barrera de la Torre 2018). Geographical knowledge production can decolonize by engaging with such border landscapes and subjectivities.

The Earth is now increasingly threatened by human resource use, exceeding what the planet can regenerate in a year. With worsening climate and biodiversity crises, geographers' discussions about the future of the *Earth* have never been so urgent. However, given the discipline's complicity with modernity-coloniality, it is an opportune moment to

---

[5] Since the 1980s, the British-Guyanese artist and photographer Ingrid Pollard has challenged representations of Blackness in English rural landscapes. She notes how 'it's as if the Black experience is only lived within an urban environment: I thought I liked the Lake District where I wandered lonely as a Black face in a sea of white. A visit to the countryside is always accompanied by a feeling of unease, dread' (quoted in Kinsman 1995).

decolonize the concept. The postcolonial writer Gayatri Spivak proposed the concept of 'planetarity' to signal humanity's subjection to larger geological forces, a point later developed by Dipesh Chakrabarty (2009). Such accounts decentre humanity and foreground colonialism's central role in configuring an uneven and unequal rapacious treatment of the Earth (Jazeel 2011). Postcolonial and decolonial geographers further grapple with the enormity of the Earth's needs: 'geographical matters are, more than ever, political, futurist, embodied cries for a different relation to our oceanic planet' (Chari 2019: 95). Rethinking Earth geographies in decolonizing directions engages with subaltern thinking that envisages the Earth in resistance to colonial-modern containment and exclusion (Chari 2019). As well as Earthly power and being, decolonial geography welcomes knowledge creation *with* the Earth. For North American Indigenous scholars, for example, 'intelligence [comes] from the ground up' through place, lands, water and urban space (Daigle and Ramírez 2019: 79; Wilcock et al. 2013).

Challenging long-held scientific periodization, the concept of the *Anthropocene* refers to the geologically significant impacts of human action on the Earth's surface. Over millions of years, geological processes left a scientifically recognizable sedimentation of minerals, rock chemistry and geomorphologic features that was unaffected by humans. Yet that is changing as anthropogenic activities create epoch-changing modifications in how we think and live with the Earth. The Anthropocene describes the 'folding of human activity into earth-surface systems such that it becomes in some sense endogenous to those systems' (Noel Castree quoted in Davis and Todd 2017: 765). In the early twenty-first century, scientists including the atmospheric chemist Paul Crutzen recorded a thin yet highly idiosyncratic sedimentary layer accumulating on the planet's surface. The layer contains radioactive isotopes from nuclear fission, concrete, plastic particles and coal dust, among other elements (Crutzen and Stoermer 2000). Scientists hotly debate whether the Anthropocene begins in human prehistory, with early agriculture (10,000 BP), or in the mid-twentieth century when such detritus rapidly expanded. These debates raise consequential questions for physical and human geography.

Decolonizing reframes the Anthropocene by making visible the coloniality of knowledge, power and being that engraves human-causes on the Earth and underpins standard accounts. Accounts query whether the Anthropocene reflects the universal impacts of all of humanity or of 'a few privileged human beings' (Luke 2018: 129) who exploit resources and territories worldwide. A universal Anthropocene narrative downplays the highly unequal privileges of a carbon-loaded mobile capitalism. Some link the Anthropocene to the colonial-modern timeline, specifically the early seventeenth-century genocides of American Indigenous populations and the subsequent environmental destruction, extraction, urbanization and ramping up of colonial capitalist earth-surface transformations (Davis and Todd 2017). Cumulatively, these processes have left earth-wide stratigraphic impacts in their wake. While this argument requires further unpacking and data (Koch et al. 2019), Anthropocene studies need to engage with colonial-modernity analyses of unequal power. One-world narratives about the Anthropocene work to hide the sharp differentiations caused by racialization, economic (in)security and reliance on climate-change inducing technologies. The power-drenched and unequal distribution of anthropogenic causes (northern, consumerist, first world) and effects (disproportionately hitting low-income groups and the global South) means that marginalized groups live lightly, consume little and use local energy sources.

Reframing the Anthropocene in relation to multiple perspectives and variable responsibilities works to decolonize explanatory frameworks, and uncovers social-spatial relations that nurture low-carbon life. Geographers and those working with marginalized groups seek 'analyses and proposals that ... reveal the inner workings of the one-world world so as to prevent their destructive capacity' (Blaser and de la Cadena 2018: 15). Plural knowledges about how to live on the Earth in non-destructive ways reveal complex relations of care and abundance, which frequently challenge the western binary of human/more-than-human. Indigenous peoples highlight these possibilities. In Arnhem Land, Australia, Aboriginal geographers stress that 'while climate change may appear as an ... unbounded phenomenon, *gurrutu* will tell us it is only ever manifested in grounded ways ... that are themselves linked to enduring place-based patterns of kinship and responsibility' (Bawaka Country et al. 2016a: 469). In low-carbon life, societies have to build multispecies abundance (Collard et al. 2015). For example, the Via Campesina transnational alliance of small and landless farmers campaign for food production with local seed banks, dignity and social justice.

Decolonial geography entails looking critically at key concepts from landscape through to Earth and the Anthropocene, reframing them in relation to geo-histories of coloniality-modernity, and acknowledging plural and decentred knowledges and relations.

IV Nature and the More-Than-Human

For many geographers, among the concepts foundational to Anglophone geography, *nature* is primary. Yet nature is a complex and slippery concept as it is a commonplace term and also unifies knowledge production in agriculture, science, conservation and other domains. Historically, western thought and practice worked with a sharp dichotomy of society versus 'nature', conceived as a 'pure, singular and stable domain removed from and defined in relation to urban, industrial society' (Lorimer 2012: 593). Yet the dynamics in this binary were always complex, as historical and critical geographers have extensively documented. Although 'imperialism [was] always an ecological project, in which humans, plants and other species were shifted around the earth in schemes for colonization/conservation' (Sidaway et al. 2014: 5), the interrelations between people, places and nature in coloniality-modernity are broader and more multi-layered than the nature–culture binary suggests.

As discussed in section 4.II, coloniality constituted racial difference by defining it in relation to an uncontrollable non-human nature. Such social distinctions emerged at the same time that European expeditions were seeking out samples of nature and accounts of human–nature relations across the globe for capitalist advantage, while museums and amateurs gathered, documented and catalogued. Neil Smith (2008) argues that the capitalist social production of nature drove material transformations of the non-human world (see also Castree et al. 2013: 335–7). From a decolonizing perspective, these capitalist–nature relations were always and irredeemably colonial–modern relations, premised on notions of supremacy and diffusionism (**Chapter 1.IV**). European manipulations, clearances, exploitation and understandings of nature – such as tropical nature – characterized and reaffirmed colonial difference.

Recently, geographers have gone beyond the society–nature binary to examine hybrid and relational interactions between entities. Human agency – and, by extension, cultural constructions and capitalist productions of nature – becomes one influence among many that combine to make 'natures'. Drawing on actor–network theory (ANT, which suggests the world comprises constantly shifting social and natural relations without independent social power), *more-than-human* geographies track the contingent networks of non-human entities such as machines, vitamins, animals and trees all networked with humans (Whatmore 2006; Lorimer 2012). This situates humans in relation to entangled processes of life and interdependence: '[humans] are not only biophysical entities made of flesh and blood (like other living creatures), but remain reliant on and daily affected by a plethora of non-human substances and actors (e.g. metals, domestic pets, and water)' (Castree et al. 2013: 388). In these 'posthumanist' frameworks, humans are not above but *in* nature, with profound implications for environmental ethics and geographical concepts of space, place and territory.

However, Indigenous and decolonial geographers have criticized posthumanist and more-than-human geographies for their ingrained colonial-modern assumptions. Indigenous geographers note the tightly-knit coexistence of Indigenous peoples with plural substances, actors and diverse biophysical beings over centuries; they also condemn the material, epistemic and societal disruptions of those relations caused by coloniality over the last half-millennium (Jackson 2014). The categorical separation of nature and society is not universal, and reflects Eurocentric 'one-world world' presumptions. In its reliance on Euro-American theory and conversations, geographical posthumanism reinforces colonial-modern hierarchies between 'modern, secular, well-educated' academics and supposedly primitive others (Sundberg 2014: 37). By contrast, decolonial geographies take seriously the site-specific dynamics through which people connect and become entangled with multifarious entities. A Māori sense of human self is inseparable from 'natural' and 'social' beings and 'connections between places, people, animals, plants, stars and gods back to the beginning' (Waitangi Treaty 2004, quoted in Johnson and Murton 2007: 125). Diverse groups' practices and epistemologies have, to varying degrees and with complex consequences, been profoundly reshaped and re-signified under coloniality. In this sense, 'colonial relationships have worked and continue to work to exclude *already existing* relational ethics' (Thomas 2015: 978, original emphasis). Decolonizing posthumanist geographies entails foregrounding Indigenous and subaltern epistemologies, and taking responsibility for the worlds we live in (Loftus 2019). In Aotearoa-New Zealand, Ngāi Tahu Māori relations with the more-than-human Hurunui River were brought into formal governance, thereby profoundly challenging Eurocentric expectations of indigeneity and nature (**Box 4.4**).

Notwithstanding the difficulty of decolonizing more-than-human geographies, constructive ways forward are being found. Yolŋu Aboriginal and non-Aboriginal geographers sought to engage Indigenous relational ontologies with Bawaka Country, which comprises sentient other-than-human beings and humans in northern Australia. A set of multidirectional relations of caring and influence in Yolŋu life-worlds or **ontology** establishes that each element (human, earth, wind, etc.) is constituted through relations with other elements, each having its own law and knowledge (Bawaka Country et al. 2016a, 2020). Practices that bring about the co-becoming of humans and more-than-humans are, for

Box 4.4 Rights of nature: rethinking nature as a legal being

Eurocentric notions of nature have been challenged in recent years with the recognition of nature's rights under the law. Whereas Indigenous peoples view other-than-human beings as integral to all being and decision-making, and therefore of full 'standing' relative to humans, that has not been the case in western legal systems. Western liberal law established human rights (under the UN's Universal Declaration of Human Rights, 1948), and nation-states and corporations (Gregory et al. 2009: 657), but these legal frameworks endorse human primacy over nature. Consequently, things falling under western categories of nature are not considered to be rights-holders. Yet in the early twenty-first century, 'nature' gained legally formalized inalienable rights in at least eight countries, and in a further 100 sub-national areas (Kauffman and Martin 2018).[1] In part, nature's gaining of rights arose from Indigenous political mobilization for interconnected human and more-than-human life. Ecuador and New Zealand created the rights of nature in national law and process. Ecuador's 2008 Constitution uses two terms – nature and **Pachamama** (an Indigenous concept for the complex living entity of the Earth, human and other-than-human beings) – and grants legal rights to nature to exist, regenerate life-cycles and undergo evolution (Waldmüller and Rodriguez 2019). In Aotearoa-New Zealand, Māori disputes over colonial-era laws led to a river and a forest gaining Indigenous guardianship (*rangatiratanga*) and to respect for ecosystems as ancestors and living entities with intrinsic value under Waitangi Treaty provisions (Kauffman and Martin 2018).

[1] For up-to-date information, see http://harmonywithnatureun.org/rightsofnature.

Figure 4.2 Te Awa a Whanganui, Aotearoa-New Zealand
Source: Yvonne Underhill-Sem

Nevertheless, rights for nature remain entangled within durable relations of coloniality-modernity. Reflecting place-specific understandings of human–nature relations, the rights of nature are difficult to establish as a singular universal model. In the United States, rights of nature are frequently contained at the municipal government scale, restricting action at the

Figure 4.3 Lake Waikaremoana, Te Uruewa, Aotearoa-New Zealand
Source: Yvonne Underhill-Sem

wider scales of human exploitation and degradation of nature. The *geographies* shaping where and how rights for more-than-humans are granted and realized thus remain a crucial area for further enquiry.

them, embedded in laws and kinship relations. Co-becoming does not exist outside the geo-histories of colonial-modern **neoliberalism** that damage these world-making practices.

One damaging colonial-modern dynamic is the appropriation of Indigenous knowledge of more-than-human relations by geographical scholarship. Knowledges of nature and the more-than-human reflect longstanding disparities in the gathering of, control over and sharing of knowledge (Loftus 2019). When non-Indigenous scholars cite non-universal epistemologies around nature this does not necessarily reflect decolonization. The Indigenous feminist Zoe Todd (2016) argues that posthumanist geographies' search for Other ways of being with nature risks (re-)colonizing, as colonial-modern epistemologies remain unquestioned. Likewise, the geographer Sharlene Mollett calls out the 'tendency for "more than human" geography to ignore both race and blackness' and to perpetuate universal notions of the human

(Mollett 2017: 13). These important points reflect geography's unexamined whiteness, and the marginalization of Indigenous and Black voices in the discipline (**Chapter 1.V; Chapter 3.IV**). Addressing this suggests that all geographers need to reflect on how their situation in relation to racialization, capitalism and settler colonialism impacts on their understandings of nature (Jackson 2014).

Decolonizing critiques of more-than-human and posthumanist geographies establish that geography's concept of nature does not have universal validity. Decolonial geography shows that the human and more-than-human are mutually constituted in place-specific processes of co-becoming, with dominant western conceptions recognized as one mode of world-making among plural others. The concept of the *pluriverse* captures this dynamic intermingled plurality of the multiple ways life and being are made, and refers to the totality of possible worlds (Oslender 2019). The pluriverse is a world in which many worlds fit, and hence a key decolonial concept. As the Zapatista movement in Chiapas, southern Mexico, explains: 'Many worlds are walked in the world. Many worlds are made. The world we want is a world in which many worlds fit' (quoted in Blaser and de la Cadena 2018: 1). Decolonial geographers advocate analysing and 'supporting existing worlding practices that enact worlds different to those produced by European imperialism and settler colonialism' (Collard et al. 2015: 328). The pluriverse concept signals that multiple worlds exist on their own terms in their distinctive heterogeneity (although they are partially connected with and through each other). Consistent with decoloniality, the concept entails a modest and critical recognition of plurality beyond standard disciplinary – and Euro-American – thinking. In contrast to posthumanism, the pluriverse concept acknowledges that life-worlds (of humans, non-human animals, other forms of living beings, geological and physical entities) are entangled with diverse and distinct epistemologies. Understanding the pluriverse thereby requires physical and human geographers to decentre the dominant epistemology in order to engage respectfully with those knowledges (**Box 4.5**).

Box 4.5 Pluriverse: 'a world in which many worlds fit'

In decolonial social science, the concept of the pluriverse refers to the coexistence of multiple ways of being and knowing, involving diverse human subjects and other-than-human entities. The concept has hybrid origins in grassroots social movements and decolonial theory. According to these thinkers, the pluriverse consists of heterogeneous worlds that emerge dynamically from the interrelations between different forms of being and knowledge. The pluriverse presumes and engages respectfully with multiple and overlapping 'worldings' (ways of being in the world, also called ontologies) that materialize across complexes of 'other-than-human persons' and humans (Blaser 2014). These worldings are not separate one from another, but are constantly 'coming about through negotiations, enmeshments, crossings, and interruptions' (Blaser and de la Cadena 2018: 6; see also Thomas 2015). For this reason, the pluriverse is profoundly frictioned and political, yet cannot be imagined in colonial-modern political categories. Peruvian other-than-human beings including mountains and forest animals enact worlds and resist the accelerated extraction of natural resources, yet cannot be brought into state decision-making (de la Cadena 2010). In Colombia, Afro-descendent groups, non-human animals and an 'aquatic space' of tides, rivers and rains comprise a worlding that attempts to resist rampant forest clearance, palm oil plantations and racism (Oslender 2019). The pluriverse is a critical decolonizing concept for understanding colonial-modern borders and contestations.

In the process of decolonizing geography, the foundational concept of nature has been profoundly rethought, as Eurocentric catalogues of natural and human domains are overturned by the inclusion of the other-than-human beings and, most recently, the pluriverse. Decolonial critiques press posthumanism to engage in an anti-colonial geopolitics and body-politics of knowledge, thereby challenging geography's whiteness and modern politics.

V Chapter Summary

The word 'geography' means writing ('-graphy') the world ('geo-'), so decolonizing geography entails changing the words and meanings we use to write about the colonial-modern world. Rather than writing only within or about the one-world world, decolonial geographies engage with and understand multiple ways of experiencing and conceiving worlds. Decolonizing geography involves changing how we write about multiple geo-socio-bio-more-than-human worlds, and whose insights and concepts we use. Geography has to delink from a longstanding set of foundational concepts which rely heavily – if largely unconsciously – on western and Eurocentric assumptions and meanings that perpetuate modern-colonial exclusions. This chapter has argued for an overhaul of geography's classic categories and concerns, many of which have been central to the discipline for over a hundred years. By decolonizing, geography moves towards a critical re-signifying, re-contextualizing and re-defining of disciplinary topics and vocabulary. Geography's decolonial turn has reoriented the discipline and transformed how it understands itself in relation to colonial-modern geopolitics and body-politics of knowledge (**Chapter 3**).

The chapter showed how decolonial and critical geographers are refining and recasting core geographical concepts in new and exciting directions. Further refinements will no doubt be made over the coming years, as decolonizing debates spread across the discipline and generate innovative vocabularies. At this preliminary stage, however, pluralizing spatial analysis resituates geography within the responsibilities and ethics of the pluriverse. Yet the discussion also indicates coloniality's ongoing power to structure our thinking. While geographies of the more-than-human and the Anthropocene might be expected to be *less* colonial-modern (because they developed in the context of critical postcolonial work), in effect they still *re-inscribe* one-world and Eurocentric assumptions. The durability of colonial-modern lenses for viewing the world reinforces the need for decoloniality to become mainstream across geography as a whole. The decolonial redefinition of core concepts – and the addition of new terms – allows us to align geographical concepts to better evaluate the spatial dimensions of coloni-ality-modernity. The chapter ended with the concept of the pluriverse, which names and conceptualizes the simultaneous and mutually respectful-responsible relations – albeit with tensions – between humans, non-human animals, landscape features, and geological and physical entities. The pluriverse represents a world brought into being in multiple ways, inviting geography to write the world differently in decolonial 'pluri-geo-graphies'.

Further Reading and Resources

As this chapter has made clear, the process of decolo-nizing geographical concepts happens in diverse ways across multiple places and spaces. The readings and resources listed below inevitably represent only a small selection. You are encouraged to seek out blogs, websites and publications that reflect decolonial geographies where you stand.

Readings

Antipode Editorial Collective (ed.) 2019. *Key Words in Radical Geography: Antipode at 50*. London, Wiley.

Davis, H. and Todd, Z. 2017. On the importance of a date, or decolonizing the Anthropocene. *ACME* 16(4): 761–80.

Wilcock, D., Brierley, G. and Howitt, R. 2013. Ethno-geomorphology. *Progress in Physical Geography* 37(5): 573–600.

Dictionaries of human geography

Castree, N., R. Kitchen and Rogers, A. (eds) 2013. *Oxford Dictionary of Human Geography*. Oxford, Oxford University Press.
Gregory, D. et al. (eds) 2009. *The Dictionary of Human Geography*. Oxford, Wiley-Blackwell.
McDowell, L. and Sharp, J.P. (eds) 1999. *A Feminist Glossary of Human Geography*. London, Arnold.

Websites

Decoloniality London
https://www.decolonialitylondon.org
'We are a London-based network committed to replacing the modern/colonial world system with justice.'

Kollectiv Orangotango+
https://www.transcript-verlag.de/shopMedia/openaccess/pdf/oa9783839445198.pdf
This is Not an Atlas: A Global Collection of Counter-Cartographies.

Southern Perspectives
https://southernperspectives.net
'The aim of this site is to promote a south-south dialogue of ideas. It profiles individuals and organizations that explore a southern perspective on a broad range of disciplines, including creative arts, humanities, professions, social and physical sciences.'

Women's Environmental Alliance and Native Youth Sexual Health Network
http://landbodydefense.org/uploads/files/VLVBReportToolkit2016.pdf
Violence On the Land, Violence On Our Bodies: Building an Indigenous Response to Environmental Violence.

–5–
Decolonizing Geography's Curriculum

In the United Kingdom and South Africa, the call to decolonize geography is associated with the student activism to decolonize the curriculum led by South Africa's Rhodes Must Fall movement. Quickly expanded elsewhere, students and educators spoke out passionately against systematic colonialism in university disciplines, pointing out that while universities are at the forefront of new information and interpretation, their teaching materials and awards of excellence still rely on colonial-modern assumptions and exclusions. In contrast, decolonizing the curriculum entails an informed reorienting of institutions, curricula and pedagogies. In geography, decolonizing the curriculum reframes how the subject is understood and broadens teaching to encompass a range of knowledges. For instance, Oxford University's Rhodes Must Fall movement advocated learning from Black consciousness, Pan-Africanism and Black feminisms to make the curriculum more inclusionary and to challenge narrow conceptions (Chantiluke et al. 2018). The lessons for universities across the world are clear: include plural voices and perspectives, read authors working outside the mainstream, acknowledge the coherence and relevance of non-European epistemologies, and provide plural frameworks for interpreting the pluriversal world (de Sousa Santos 2014; Daley 2018).

Decolonizing the university does not, however, mean the wholesale replacement of western with non-western epistemologies, nor prioritizing one approach over another. Instead it means reflecting on how to make sense of the world where plurality exists always in relation to coloniality. Moreover, it involves universities in self-critical examination of academic pedagogic practice at the micro-scale (Daya 2021). With universities taking on new knowledges, decolonizing delves into how best to interact with students, cite new sources and engage with non-university teachers and instructors. This chapter is primarily addressed to instructors and students who are racialized as white and live in wealthier countries, as their geographical knowledges require decolonizing most, to relieve oppression of students and instructors of colour.

Yet educational institutions perpetuate **deep colonizing** even in the most progressive programmes (Howitt 2001b). However much a curriculum changes its authors and topics, decolonization cannot occur if deep structural processes continue to reproduce a one-world world geography, both by normalizing Eurocentrism and by excluding plural subaltern knowledge systems (**Chapter 3.I**). Over centuries, places of higher learning have been at the heart of authorizing westernizing knowledge and disassembling alternatives generation after generation. By transmitting colonial-modern attitudes and behaviours, universities restrict the range of understanding and experiencing of the world. Nevertheless, geographers and educationalists are finding ways to identify and challenge the links between knowledge, instruction and power. Critical pedagogies open up the classroom to the world and dismantle sanctioned ignorance and classroom exclusions. Practical steps encourage post-secondary geography staff and students to delink from a sole focus on dominant epistemologies. Although geography remains a deeply exclusionary discipline, the educational resources and practical tools are now available for decolonizing and more inclusive learning.

I Coloniality and Education

Education is a key site where coloniality is reproduced, as its formal purpose is to pass on knowledge from one generation to the next. The direct connection between colonialism and the discipline of geography continued well into the twentieth century, as my department at Cambridge illustrates. Recalling undergraduate geography in the mid-1940s, A.T. Grove (b. 1924) recounted:

> I think the geography department in Cambridge was unusual in offering topographic and geodetic surveying as a paper in the [final] year ... And most of the people who did that went into the Colonial Survey department, and that is what I had in mind in my third year. ... I did physical geography and geodetic and topographic surveying ... We were taught by Sir Gerard Lennox Cunningham, who I think had been director general of the survey of India around about 1912. (Grove 2010: 59)

The Colonial Survey department has long gone. Nevertheless, post-secondary education in universities and colleges plays a key role in defining (and defending) professional fields, imparting the latest research, and providing a rite of passage into adulthood. It defines what and whose knowledge counts in the world at large. Because educational systems were established in colonies to mimic and reproduce metropolitan disciplines and curricula, postcolonial countries frequently internalize coloniality. Colonial-modern forms of knowledge continue to have global reach and are endorsed by metropolitan institutions, consolidating the perception of Eurocentric disciplines and content as self-evident and useful globally (Connell 2007; Cupples and Grosfoguel 2019). For instance, the South African scholar Fana Sihlongonyane (2015) notes how the planning and geography disciplines are 'insularly European', and hence devalue African realities and knowledges. **Westernizing universities** do not dismiss local realities and knowledges in an incidental or haphazard way.

Rather, knowledges are selectively reinforced, passed over or lauded, with profound implications for lives beyond campus as they influence material and political, economic and psychic relations.

A discipline such as geography has core reference books and authors that are generally agreed to represent the standard against which new knowledge is evaluated, and which are overwhelmingly western-informed and Euro-American-centred. Through unquestioned criteria of selection, the **canon** defines the norms by which other knowledges are judged to be inferior, unreliable or potential additions to the canon. Since the late eighteenth century, western designs have broadly set the structures of knowledge (in terms of academic disciplines) and defined what counts as valid knowledge (Mignolo 2002), following the Enlightenment's separation of modern knowledge from 'tradition' (**Chapter 1.IV**). Colonial ideologies justified campaigns against what was termed idolatry, ignorance and tradition. The sixteenth century in particular saw a violent intolerance towards the thought, experience and learning of colonized populations. Whether in the Catholic Iberian re-conquest of Moorish Spain in which Islamic and Jewish libraries were burned, the destruction of native American written texts, or the disarticulation of African societies through enslavement (Grosfoguel 2013), educational institutions were built in their ruins. Coloniality thereby pursued 'the destruction of the knowledge and cultures of [colonized] populations, of their memories and ancestral links and their manner of relating to others and to nature' (de Sousa Santos 2016: 18). That **epistemicide** continued through the nineteenth and twentieth centuries. For example, European settlers in Chile, Canada and Australia forcibly separated native children from community knowledge-transfer and placed them in residential boarding schools to teach them European languages, disciplines and behaviours (Wahlquist 2016). Whether in Europe, postcolonial countries or settler colonies, colonial-modern explanations 'trivialize, distort, misunderstand, misuse, romanticize [non-dominant] knowledge and systems of thought' (Nakata et al. 2012: 128).

But – we might ask – isn't this epistemicide over today, now that the colonial residential schools are closed and intercultural education is in place? By no means. Geographers report how school and college teaching resources uphold colonial understandings of the world, and block understanding of a range of subaltern experiences. When Canadian geography instructors discuss the 1876 Indian Act as an historical event, the law's *current* effects are not considered, despite Indigenous knowledge of its ongoing influences on territory and citizenship (Godlewska et al. 2010). In this and other cases, the curriculum is made authoritative by not recording and listening to subaltern knowledges (Kuokkanen 2008: 62; **Chapter 3.I**). Anglophone geography inducts students into a selective body of knowledge, which combines a Eurocentric-authorized curriculum with sanctioned ignorance – both outcomes of historic violence (De Lissovoy 2010: 285; Tuck et al. 2014: 1; UCL Collective 2015). These characteristics in turn shape geography's 'citations, curricula, canons and recruitment patterns' (Elliott-Cooper 2017: 333).

Coloniality confirmed a sanctioned body of work known as a canon, that is, an established and valued collection of writings, cultural artefacts and artworks associated with a country or discipline. The western canon is a set of literatures and artworks deemed to be classics in metropolitan countries, setting the standard by which to judge other knowledges, writings and arts. Colonial forms of knowledge deeply influence the geographical canon in universities worldwide. In India for instance, only one fifth of the authors cited in urban geography reading lists are Indian, and one tenth in political geography (Sundaresan 2020). By constructing a canon, Eurocentric worldviews, core texts and theoretical frameworks become 'geography', due to their presentation as comprehensively reflecting the world. Although an author's nationality does not determine viewpoints, it does shape the topics, content, interpretations and examples selected. 'Decolonization asks us to consider how the location and identity of an author shape their perspective. … Which [authors] are privileged and placed at the centre? Whose voices are authoritative and considered part of the canon while others are left at the margins?' (Muldoon 2019). For this reason, the postcolonial critic Sara Ahmed calls the canon a 'reproductive technology, a way of reproducing the world around certain bodies' (cited in Ybarra 2019). Confronting the educational structures of coloniality thus involves questioning the canon and the wider dynamics of teaching and learning for multiracial, multicultural students. Citing alternative sources and knowledges contributes to a decolonizing re-claiming of knowledges viewed as open, varied and inclusive (Scott Lewis 2018).

Beyond changes in curricular content and purpose, decolonizing has to involve broader commitments to transforming the institutions of learning. Leadership at various levels needs to engage in profound reform and facilitate changes in teaching and staffing. Such changes are obstructed by racism in hiring practices and by neoliberalizing tendencies in university priorities. Universities reliant on student fees and corporate sponsorship may be less willing to engage in critical discussions, preferring to increase class sizes and develop status-quo-endorsed vocational courses. Under neoliberalization, universities are pressured to not challenge disciplinary boundaries, faculty hiring and pedagogic innovations (such as community involvement). In this regard, universities 'do the job of deep-colonizing' (McLean et al. 2019: 124; Howitt 2001b). Despite appearing to reverse coloniality, in practice they risk reproducing it. To take one example, the welcome inquiry into the University of Cambridge's benefits from and challenges to trans-Atlantic enslavement carries a 'danger that the university may seek to excuse itself from enacting action in the present by directing general focus to its deplorable presence in the past' (Davis 2019). Acting in the present means decolonizing the curriculum and overcoming racism and intersectional exclusions for staff and students.

To counter co-optation and deep colonizing, critical geographers and educationalists point to university geography's potential for inclusive and plural ways of knowing and being as it begins to 'unsettle colonial and racist knowledge production and systematic oppression in and beyond the academy' (Daigle and Sundberg 2017: 338).

II Decolonizing the Teaching-Learning Process

For critical educators, teaching and learning are dynamic processes where power relations are challenged and new thinking can flourish; the classroom thus becomes a radical space of possibility for transforming interpersonal relations and mutual learning (Icaza and de Jong 2018; Hinton and Ono-George 2019). In geography this encourages learning about diverse world-writings and lives in the world, while challenging micro-incivilities and the discipline's white norms in the classroom. To contextualize discussions coming later in this chapter, this section introduces the principles of decolonial and critical education, and three transformations in particular:

- Expanding the scope of learning
- Bringing coloniality-modernity into explanatory frameworks
- Critical (self-)reflection on social positionality

Together, these elements set the direction for changes to curricula and racialized classroom interactions, to construct pedagogies of solidarity and expand learning beyond the classroom.

Education, according to decolonial educationalists, provides an arena where sanctioned epistemic ignorance can be transformed in dialogue and learning among students and instructors. The Brazilian philosopher Paulo Freire outlined a world-changing 'pedagogy of the oppressed' that critiqued classic education as a form of 'banking' where teachers transmit canonical knowledge in pre-defined packages to students' memory (filling student memory banks!). By contrast, Freire's pedagogy of the oppressed began with the idea that education allows the oppressed to regain a sense of humanity and become critical and aware learners (Freire 1971). Extending these objectives, decolonial educationalists foster 'learning dispositions in students that encourage openness to further enquiry' (Nakata et al. 2012: 120). In a two-way process between learning-students and learning-teachers, decolonizing decentres teachers' authority, diversifies learning materials, and values plural experiences and knowledges (De Lissovoy 2010; McLean et al. 2019). In order to situate varied knowledges geo-historically, western and plural others are each identified as such. Geography moves towards dialogue – sometimes difficult and tensioned – where no monopoly on knowledge exists.

Decolonizing 'starts from epistemic de-linking: from acts of **epistemic disobedience**' (Mignolo 2009: 15), in order to shift how we understand and treat the world. Working with that insight, decolonizing extends our understanding of what counts as 'reason beyond Eurocentric and provincial horizons' (Maldonado-Torres 2011: 10) to appreciate plural perspectives. The expertise of diverse Aboriginal, BAME, Latinx, Black and other knowledges embed a curriculum in a less-Eurocentric worldview and enhance the quality of geographical education. Decolonizing hence does not abolish the western canon entirely but reframes its sum of knowledge as one element among plural others. Decolonial curricula are of course highly varied, as content and priorities will vary with context (Daigle and Sundberg 2017). African universities assign readings on Pan-Africanism, nineteenth-century African-American politics, civil rights movements and Afrocentricity (Ndlovu-Gatsheni 2013; Langdon 2013; Katundu 2020), while in North America, curricula juxtapose Indigenous and canonical authors (**Box 5.1**).

Decolonizing teaching-learning also involves critical (self-) reflection on social positionality, especially among those perspectives that are solely informed by colonial-modern education and knowledge-holders. Recognizing that 'universal' knowledge systems block an understanding of other sources and interpretations is a key step in decolonizing education. University-level geography can work towards decolonizing by firmly situating its 'one-world world' knowledge (**Chapter 1.VI**) in relation to coloniality-modernity and critical alternatives (Esson 2018). Colonial-modern knowledge dynamics silence, exhaust and alienate non-white and subaltern students (Noxolo et al. 2012; Langdon 2013). Decolonizing geography's curriculum thus involves much more than changing

Box 5.1 The challenges of decolonizing a university

Cases from Mauritius, Australia and Nicaragua illustrate how university efforts to decolonize the curriculum can be hampered by processes of 'deep colonizing'. In Mauritius, social science scholars outlined an ambitious agenda to decolonize their university, taking into account meagre resources, structural inequalities, colonial histories and an international student body (Auerbach 2018). The ambition was to build a new canon and ways of knowing. Among the decolonizing elements were:

- Students and instructors work together on decolonization
- Students engage with diverse non-textual materials (e.g. artefacts, music, advertising, architecture, food)
- Students contribute to changing public discourse around Africa, by producing op-eds, blogs and vlogs
- Use open source materials
- Read at least one non-English text a week
- 1:1 ratio for student exchanges with western universities
- Teach and maintain high ethical standards

However these decolonial ambitions were cut short when university authorities and a private corporation replaced these measures with neoliberal curricula and European accreditation. Decolonial advocates resigned in protest. The Mauritian example highlights the contested nature of the ongoing colonial power at stake in higher education.

In Australia, the University of Adelaide devised a three-year experiment to incorporate Indigenous knowledges into a range of courses. An Indigenous Reference Group was formed to agree underlying principles of curricular reform. The group focused on challenging essentialized representations of Indigenous groups and highlighted the spatial diversity of Indigenous lives in dynamic landscapes (Nursey-Bray 2019). In the classroom, settler and Aboriginal teachers used dialogue and diverse materials to disrupt unspoken narratives. Indigenous knowledge was incorporated so learners engaged with Aboriginal senses of 'Country'. First-year students were tasked with researching an Australian city in relation to the relevant Indigenous Country and its experiences since colonization. Yet Aboriginal knowledges were not given parity with Euro-American frameworks and the institution remained racialized and westernizing, thereby perpetuating deep colonizing (Nakata et al. 2012). Taking decolonizing further would begin with Indigenous leadership and governance. At the university in Nicaragua's Caribbean coastal autonomous region, instructors worked together with Indigenous and non-Indigenous students and intellectuals to overcome racism and sanctioned ignorance (Cupples and Glynn 2014).

reading lists, since it must also address structural problems and nurture justice-building, empathy and non-defensive self-reflection (De Lissovoy 2010: 282).

III Decolonizing the Curriculum

Educationalists suggest that decolonizing curricula works against domination because it makes visible the Eurocentric epistemic and cultural violence that underlies educational content and structures (De Lissovoy 2010). A decolonizing curriculum encourages inquiry, self-reflection and critical analysis, laying the foundation for learning after university education. So how does a geographer go about decolonizing a curriculum?

Decolonizing the canon

Anglophone geography tends to view itself as a discipline without a rigid canon, especially in comparison with other subjects. Since human geography's cultural turn, English-speaking geographers have focused on power, discourse and difference (Blomley 2006: 89). In comparison with political science or anthropology, post-secondary geography sees itself as unfettered by nineteenth-century heroes and takes pride in its interdisciplinarity and paradigmatic openness. These scholarly characteristics, however, coexist with a tendency to reference and value scholarship embedded in north Atlantic (and Anglophone) institutions that sideline discussions of racialization and Other knowledges (Esson 2018; Carter and Hollinsworth 2017). Reading lists are often overwhelmingly white in geography and other disciplines (Peters 2015). Compounding the selectivity is the widespread use of canonical European social theory from Marx to Foucault. Familiarity with these references and theories becomes the narrow benchmark of academic excellence for post-secondary students (**Chapter 3.IV**). By contrast, decolonizing changes the range and type of thinking and information presented to and discussed with students. The goal is to change the conversation, not merely 'add Black and Indigenous authors and stir' (Shilliam 2015; Rutazibwa 2018).

Changing these components in turn reorders the progression and the goal of decolonizing across and through geography programmes. Placing postcolonial and decolonial geographies in a stand-alone course risks sidelining decolonizing, although it does permit fuller examination of decoloniality's breadth and depth. Introducing decolonial and postcolonial material at the start of a higher education programme is useful as it generates ad hoc conversations among students and staff. For instance, a Canadian introductory human geography course aimed 'to unsettle Eurocentric colonial discourses that many students carry into the classroom' and 'lay the groundwork for future engagement with Indigenous and decolonial theory' (Daigle and Sundberg 2017: 338). Its themes included the spatial and scalar politics of coloniality, capitalism and geography, informed by Indigenous, critical race, feminist and queer writings (Daigle and Sundberg 2017: 339). Thematic courses offer opportunities to teach students about plural perspectives on one specialism. For example, an Australian resource management course taught relational approaches, various knowledge systems' approach to resources, and social justice (McLean et al. 2019: 128). Decolonizing geography has the potential to inform curricula across human and physical geography, from introductory to advanced classes.

Reading and more

Post-secondary learning in geography frequently relies on students' independent reading to prepare for and complement classroom activities. However, students from unrepresented groups often find that canonical texts are alienating and difficult, as they neither reflect nor validate their experiences (Ndlovu-Gatsheni 2013). Students then seek out additional reading, which can be a fraught and time-consuming task if libraries and teachers are unsupportive (Decolonizing the Curriculum 2017; Bhambra et al. 2018). This outcome illustrates deep-colonizing, as it burdens disadvantaged students and leaves colonial-modern curricula in place.

Crucially, delinking from colonial-modern frameworks means transforming how all student-learners read and interpret materials critically, as well as reading more widely. Engaging with plural epistemologies and topics of concern changes the approach to learning, involving a more diverse set of sources, and reflecting knowledges and experiences unfamiliar to most white, metropolitan geographers. In general, instructors and students have not encountered pluriversal and decolonizing information, arguments and vocabulary, so class discussions have a key role in facilitating understanding. Understanding relies on instructors providing information on content, context and authors. In my final-year course, students have classroom discussions about sources, and are directed to websites about cited authors, collectives and social movements (in translation). Students are encouraged to read and reflect on sources in relation to the politics of knowledge production (**Chapter 3**), and to identify counterpoints to Eurocentric thinking.

Reading by itself is not decolonizing unless instructors spend time explaining how to *interpret* texts in relation to coloniality-modernity and plural experiences, and draw out anti-colonial and decolonial interpretations (Dorries and Ruddick 2018). Such learning objectives are furthered by reading diverse sources in juxtaposition, to highlight the partiality of canonical sources and the situatedness of alternatives. The First Nation scholar Audra Simpson, for example, puts western authors next to Indigenous scholars for class discussion. Likewise, reading President Truman's canonical 1949 speech on underdeveloped nations alongside Aimé Césaire's 'Discourse on Colonialism' challenges development geography's canon (Césaire 1972; Rutazibwa 2018). Such juxtapositions can be challenging for students. However, learning arises from discussion of each text's goals, audiences and assumptions, while recognizing the subaltern work's distinctiveness and difficulty (Castleden et al. 2013; Dorries and Ruddick 2018).

Just as decolonizing involves engaging with sources and types of experience and knowledge outside the colonial-modern curriculum, so too it means using and learning from sources other than academic books and articles. Decolonial materials can include varied genres, such as podcasts, documentary films and talking with people with oral knowledges. Documentaries from different knowledge-holders provide insights into self-empowerment and the place-specific contexts of knowledge production. Engaging with land and its embodied knowledges adds a distinctive learning opportunity to most geography field trips (**section 5.VI**). With virtual learning environments, an increased range of films and podcasts can be shared. In each case, the pedagogical action of disrupting the canon lets in other ways of knowing, sensing and living, which challenges unspoken disciplinary norms and boundaries.

IV Tackling Classroom Racism

Decolonizing the teaching-learning process crucially engages racialized relations in the classroom. As the classroom expresses power relations where knowledge is produced and negotiated, the dynamic between students of colour, instructors and white students often results in non-white knowledges being marginalized or made hyper-visible. Geography's low percentages of non-white students and instructors in many settings makes racialized intra-class dynamics widespread and highly problematic in the discipline (**Chapter 1.V**). In this context, decolonizing means establishing and maintaining a shared ethical understanding among instructors and learners that racialized hierarchies impact negatively on learning and social relations. Decolonizing places responsibility on institutions and trained teachers to enact anti-racist policies and to hold awkward conversations about racism as a lived and institutionalized phenomenon.

Black and Indigenous people and people of colour are underrepresented in geography teaching staff and universities in the Anglophone world, reducing the visibility of role models for BIPOC students. Asking 'Why isn't my professor black?' became a catalyst for anti-racist student campaigning at UK universities in 2014 (Tate and Bagguley 2019).[1] To counter this exclusion, Black geographers organize online platforms for networking between students, graduates and academic and other professionals.[2] Such solidarity is vitally needed in light of the pervading anti-indigeneity and anti-Blackness in education (Catungal 2019). In too many institutions,

---

[1] BAME faculty face structural inequalities at universities in the UK. In a survey, Black and Minority Ethnic (**BME**) academic staff were found to have a 9 per cent pay gap with white staff averages; over two-fifths of BME staff were on insecure, fixed-term contracts (compared with one-third of white staff). Intersectional gender disparities were notable: around a quarter of white male academics had fixed-term contracts, compared with nearly half (45 per cent) of Asian female academics (information from University and College Union UK, 25 February 2020).

[2] These include Black Geographers in the UK (blackgeographers.com) and the Black Geographic in the USA (https://www.theblackgeographic.com).

drawing attention to racism is interpreted negatively rather than as a step towards anti-racism (Ahmed 2012). University classrooms involve continual relational negotiations over racial difference, when students and teachers of colour are thought to be biased or incompetent for expressing a truth that is not widely acknowledged (Mahtani 2014; Chantiluke et al. 2018). In Australia, Indigenous student participation rates across all subjects increased between 2006 and 2011, yet did not reach levels proportionate to their population (McLean et al. 2019: 127). In these contexts, Indigenous, Black, Latinx and other marginalized students experience university education as 'isolating and alienating, as they are often not taught by Indigenous [or other] scholars' (McLean et al. 2019: 127). Across tertiary education, BAME and Indigenous students are treated as hyper-visible minorities, although experiences vary across institutions. In Canada, First Nation and Indigenous students reported systematic discrimination in student–instructor and student peer-to-peer interactions, reducing their sense of security and belonging on campus.[3]

How can anti-racism be furthered in post-secondary settings for learning-teachers and learning-students? Geography instructors receive professional training in teaching, yet rarely is anti-racist pedagogy an integral element (**Box 5.2**). Tackling racism is less about highlighting individual shortcomings and more about recognizing racism in shared thinking, attitudes and procedures (Gillborn 2006). Teaching involves 'begin[ning] where students are, creating space and encouraging them to share and reflect on their own understandings and experiences of "race" and racism' (Hinton and Ono-George 2019: 2). Learning that everyone is complicit in racism, and that everyone has agency and responsibility to change that, is a key step (Tate and Bagguley 2019: 8). A discussion of complicity, however, discomforts some students (Motta 2018). Calling out anti-Black and anti-Indigenous behaviours and views is important, as is the strategic use of white instructors' 'visibility and social audibility' to enforce

[3] See https://intheclass.arts.ubc.ca.

> ## Box 5.2 Critical race topics and geography
>
> Human geography's canon does not include critical race theory. Despite its relevance for a wide range of geography topics, critical race theory tends not to be systematically taught or discussed with students (Esson 2018). To overcome geography's deep-seated tendency to teach about white people's experiences rather than those of people of colour, the historical geographer Mona Domosh (2015) proposes teaching a range of geographical themes through the lens of critical race theory, including:
>
> * Race and space in maintaining structures of domination, subordination and inequality
> * Intersectionality and space
> * Ideology of white supremacy and the use of space to maintain it
> * Critical race theory and spatialities of white privilege
> * Racial residential segregation and inequality
> * Racialization of immigrants of colour
> * Environmental racism
>
> This list's detail and breadth highlight the urgency of decolonizing with and through critical race theory.

anti-racism. Anti-racism aims to nurture solidarity with non-white students and teachers, and re-signify experiences of exclusion and violence (Daley 2018).

As most geography staff are racialized as white, educationalists and geographers agree that directly teaching about whiteness is crucial. One Australian teaching team worked by 'positioning ourselves and most students as settlers … [to] critically engage with white privilege and neocolonial structures' (Carter and Hollinsworth 2017: 193). Unsettling whiteness involves self-reflection and 'unlearning'. Resituating white authority in the classroom also happens when teaching staff of colour provide instruction, thereby decentring white authority and contributing to plural knowledge transmission (McLean et al. 2019). In North America and Australia, geography professors invite Indigenous and marginalized groups to teach, after preparing students with readings, tutorial discussions and in-class writing exercises (Daigle and Sundberg 2017: 339). Nevertheless, neoliberal, budget-constrained universities may be unable to support such additional teaching.

In sum, decolonizing the curriculum requires deliberate anti-racism interventions to call out racialized dynamics in the classroom and in the university more widely. Reshaping the dynamic between whiteness and classroom instruction, training teachers in anti-racism and calling out racism contribute to a less exclusionary learning experience for people of colour.

V Decolonizing Pedagogies

Pedagogy refers to an approach to teaching, and considers how and through what means learning takes place. Decolonizing pedagogy is an ongoing process rather than a one-class, stand-alone component. It works best when integrated across and through a programme to build students' depth of under-standing. Decolonizing geography encourages teaching to prioritize critical perspectives on coloniality, and include topics and theory that highlight the coherence and speci-ficity of non-western epistemologies. Decolonial pedagogy additionally treats the classroom as an arena for learning about respectful and dignified consideration of each other's experiences and histories (Freire 1971). Decolonization seeks to encourage learning through solidarity and horizontal relations in interactions between students, and between students and teachers (Icaza and de Jong 2018). A pedagogy of solidarity involves encouraging reciprocity, facilitating encounters between different experiences, and embedding routine practices of solidarity (Gaztambide-Fernández 2012). However, rather than offering an off-the-shelf model, decolonizing pedagogy encourages adjustments to local context, reflecting the specificity of an instructor's classroom space, context and associated responsibilities.[4]

In the classroom, instructor-led ground rules establish the type and tone of interactions between student learners in order to build self-reflective and mutually inclusive relations. Teachers may need to repeat and clarify the ground rules, which can be designed and agreed with class participants (Cook 2000). These rules can include listening to a variety of perspectives, maintaining confidentiality outside the classroom, talking about themes rather than characters, and recognizing that opinions expressed in the class are not representative of all experiences. Nevertheless, difficult conversations continue to occur between decolonizing teachers and students, which might offer a 'welcome opportunity to reinforce ... classroom agreements' rather than blame individuals (Hinton and Ono-George 2019: 3–4). In such dynamics, the class size and mode of delivery (large lecture, seminar or small group discussion) matters, as it influences the baseline interactions to be maintained. Smaller classes and seminars permit students to listen and interact more with peers, while larger classes hinder efforts to build dialogue and trust.

A core principle underlying decolonial pedagogies is to ensure student-learners understand how and why complex and contested knowledge relations exist. Teachers thus 'stress the legacy of a very complex and historically-layered contemporary knowledge space' (Nakata et al. 2012: 132). In part, this becomes a discussion around the historical and conceptual factors that limit an individual's understanding and disciplinary knowledge (Nakata et al. 2012). In this way, an acknowledgement of individual positionality can be a step towards learning about broader patterns of the coloniality of knowledge (Catungal 2019). Dialogue between teachers and students negotiates between and within multiple positions. In this respect, the classroom is a window onto – and an integral part of – embodied, relational dynamics of power

---

[4] See https://intheclass.arts.ubc.ca/discussion-topics/topic-3-classroom-incidents.

and knowledge. While providing safe spaces for mutual learning, the task of mediating between racialized position-alities can be challenging for educators, who require support from university action and policies.

A key decolonizing pedagogic tool is the use of non-standard classroom interactions and assessments (Mbembe 2016). For relatively privileged university geographers, working with and for marginalized groups in and outside the classroom holds potential (Nakata et al. 2012). Laying aside essays and presen-tations encourages students to learn from alternative spaces, objectives and audiences. Narratives and storylines provide constructive starting points for these goals (Nursey-Bray 2019). The Brazilian activist Augusto Boal designed forum theatre (also called theatre of the oppressed) as a means to encourage engagement and familiarity with a scenario (a role play, theatre piece or documentary). The scenario runs through once and is then repeated so that student-learners can stop the action when they hear or see something linked to the discussion topic, such as facets of coloniality (Langdon 2013). Non-standard interactions and assessments encourage student creativity through poetry, music and sound (Gaztambide-Fernández 2012), as well as museum and gallery visits. Journal diaries encourage students to record and reflect on daily examples of coloniality or personal histories and identities (Cook 2000). Non-written assign-ments such as art installations, craft pieces and temporary public plaques around a city promote active learning through encounters with non-university knowledge-holders, non-canonical materials and embodied experiences (Howitt 2001b; **Box 5.3**). Finally, 'blended learning' combines online materials, online interactions and face-to-face classroom-based teaching, providing opportunities for decolonizing pedagogies (McLean et al. 2019).

One decolonial curriculum debate centres on teaching non-Indigenous geographers about Indigenous peoples, because of the prevalence of colonial-modern *mis*-representa-tions of the latter, rather than Indigenous self-representation. Decolonizing thus involves unlearning colonial-modern repre-sentations, critiquing their origin and power, and gaining an

Box 5.3 Global lives in public spaces: cultural and historical geography

Fierce debates around public memorials and statues have long animated decolonizing and anti-colonial agendas (**Chapter 1.I**). Echoing South Africa's Rhodes Must Fall movement, marchers for Black Lives Matter pulled down a prominent statue of the slave trader Edward Colston in June 2020 in the city of Bristol, UK. Critical geographies place these struggles in their historical, urban and public contexts. An optional second-year undergraduate course at the UK's University of Exeter explores colonialism, power and urban landscapes in the city of Exeter, in Britain's southwest. In 'Global Lives: Multicultural Geographies', students think about how traces of global flows and connections to Exeter are made public or invisible in the urban landscape and in museum collections. The city's diverse voices and artefacts are accessed through walking tours, led by the 'Telling Our Stories, Finding Our Roots' community group. Student research is undertaken using postcolonial and decolonial geography and museum studies, supported by the Devon and Exeter Institution, the 'Legacies of Slavery in Devon' group and the city's Royal Albert Memorial Museum (RAMM). Drawing on the guerrilla memorialization literature (Rice 2012), each student produces and temporarily places a new 'blue plaque' in the city to highlight hidden connections, or creates a decolonizing intervention for an item exhibited in the RAMM. Reflection on whiteness and colonialism occurs throughout the course, informed by anti-racist and Freirean pedagogies, addressing Anglophone geography's unspoken whiteness (**Chapter 1.V**) and compensating for British school-leavers' lack of background knowledge on colonialism.

understanding of diverse Indigenous positionalities in/at the margins and against coloniality-modernity. Significantly this queries the western binaries between non-Indigenous and Indigenous that block decolonial understandings of subaltern heterogeneity and complex positionings (Nakata et al. 2012: 134; Radcliffe 2017b). Decolonial geographies 'deliberately uncouple the binary between dominant constructions about indigeneity and places, and the normative values ... ascribed to Indigenous knowledge and culture' (Nursey-Bray 2019: 327–8; **Box 5.4**). Learners come to understand the importance of disaggregating, situating and contextualizing Indigenous peoples, and welcoming heterogeneous accounts from them (Nakata et al. 2012).

Decolonizing pedagogies aim to nurture a critical questioning of dominant ideas and to extend solidarity (De Lissovoy 2010). They question monolithic categories, provide non-essentialized interpretations of relational processes, and recognize and work with intersectional positionalities.

VI Opening the Classroom to the World

As a fieldwork-based discipline, geography is committed to expanding understanding of the world through students' direct learning from embodied encounters with people, environments and places outside the classroom. What contributes to decolonizing teaching and learning in the field? Since decolonial scholars suggest that routine processes in everyday life are imbued with coloniality, fieldwork aiming to engage with coloniality can potentially take place anywhere. After classroom preparation, field visits have the potential to decolonize geography in four main ways:

• Students engage with non-academic communities and knowledge systems
• Students and teachers decentre canonical text-based knowledges
• Students learn about how and why the physical and social features of a place relate to coloniality

• Teachers and students decentre divisions between physical and human geography by placing them in wider frames of reference

Decolonizing transforms the purpose and goals of extra-classroom geography in order to unsettle student expectations and epistemologies (Abbott 2006). It offers the opportunity for meeting knowledge-holders whose information and positionality differ from that of university instructors. Moving off-campus and engaging with other modes of learning can provide a corrective to classroom and canonical content. One Australian university, for instance, deliberately reduced classroom-based teaching to spend more time learning 'with Country' from Aboriginal elders (McLean et al. 2019).

In general, working outside the classroom vividly highlights the material and embodied dimensions of the colonial present. Social justice dimensions of decolonial geography can be explained in face-to-face interactions with non-university knowledge-holders who have experience of inequalities. Although often marked by difference, such encounters can build into long-term changes in perspective and understanding after classroom discussions about intersectionality, whiteness and racial hierarchies. For instance, the 'Grow Dat' Farm near New Orleans instructs youth and adults about slavery, segregation and privatization. Using Freirean principles, its lesson is that 'land itself can hold history [and geography] in a different way to a textbook' (Brown et al. 2020). To access historical geographies, documentaries can provide key points. The short film, 'The lost neighbourhood under NY's Central Park', for example, documents how the African-American rural village of Seneca was displaced to make way for this urban park.[5]

Decolonizing the teaching about physical landscape features can begin by referring to how visible contemporary

---

[5] See   https://www.vox.com/2020/1/20/21070883/central-park-seneca-village.

landscapes have been worked over by historical colonialism and by scientific-political projects in the colonial present. Rivers, soils, ecosystems and vegetation profiles all bear traces of colonial influences, from plantation agriculture, the clearing of people and vegetation, engineering for colonial and settler projects, to current conservation projects, and so on. Educating geographers about these processes, and offering them field-based experiences, can challenge the frameworks for understanding based in colonial-modern knowledge. In world regions where non-white spatial histories and experiences have been systematically erased from the landscape, it is particularly valuable to seek out information on Black, Indigenous, Aboriginal and other groups' presence and landscape impacts (Nxumalo and Ross 2019).

As well as connecting geophysical features to colonialism's ongoing effects, decolonization engages with non-university-based knowledge systems, which can occur in a number of ways. One strategy is to juxtapose physical geographical models and Indigenous classifications in discussing topics such as subterranean layers, water landscapes and landforms. Undergraduate students introduced to alternative knowledge systems gain respect and appreciation for them, especially through interacting with knowledge-holders. For example, in central Canada, diverse undergraduates immersed themselves in Indigenous plant knowledge and practices, discovering different ways to see and know ecological systems (Bartmes and Shukla 2020). The fact that fieldwork takes place in a specific location can itself help to delink learning from Eurocentric over-generalizations. Instead, students can be encouraged to consider 'diverse, specific, and non-generalizable' relationships to land and place (Tuck et al. 2014: 8–9). Decolonizing physical geography is not then about jettisoning western scientific insights, but instead brings them alongside plural legitimate knowledge systems that extend students' learning.

When done respectfully within pedagogies of solidarity, learning with non-university knowledge-holders in situ can establish learners' appreciation of plural forms of knowledge which have been absent from their school or university education. Students' direct embodied interactions with people and places can be an opportunity to unsettle the certainties of canonical education through immersion and experiential learning. Whether in field visits with immigrants, refugees or Indigenous representatives, this learning disrupts stereotypes by focusing on the coherence of in-place experiences and perspectives. With critical intersectional pedagogies, these engagements can nurture forms of 'two-eyed seeing' (Corntassel 2020; Bannister 2020). Learning two-eyed seeing is as relevant in physical geography as it is in human and environmental geographies, as it builds skills in intercultural collaboration.

In settler colonized places, geography students often reach tertiary education with a poor understanding of Indigenous and other displaced groups' knowledges. Field-based working with Indigenous teachers directly conveys that knowledge and widens students' appreciation of (depending on the context) Black, Indigenous, migrant and people of colour's learning *from* the land (L. Simpson 2014). Land-based pedagogy is decolonizing when geography students come to think of the land as the first teacher on issues of Earth stewardship and living forms of knowledge associated with other-than-human beings (Tuhiwai Smith et al. 2018; Bartmes and Shukla 2020; **Box 5.4**). In decolonial pedagogy, fieldwork raises awareness of non-academic groups' distinct epistemological positionings, and contributes to plural literacies of land across physical and human geography.

Fieldwork is widely perceived as being central to geography's ethos and analytical distinctiveness. Yet geography fieldwork too often denies racial hierarchies and colonial processes, and provides merely a 3-D ratification of textbook processes described through one-world world lenses. In decolonial pedagogy, by contrast, moving outside the classroom is an opportunity to delink from Eurocentric canons and 'place-less' science. Decolonization works with multiple knowledge systems to generate situated literacies about places.

Box 5.4 Learning from the land

If colonialism ruptured Indigenous and other lives, decolonizing aims at the reconnection of Indigenous peoples to land and its connections to life and knowing (Wildcat et al. 2014). Informed by Indigenous theories of resurgence (**Chapter 3.IV**), physical and human geography has much to learn from the land. For example, since 2010 the Dechinta Centre in Canada's Northwest Territory has run accredited programmes on plants, ecology and ethno-medicines, where Dene First Nations, University of British Columbia academics and others teach about the interlinkages between ecological, cultural and linguistic diversity (Bannister 2020).[1] Indigenous educationalists provide instruction on Indigenous knowledge of the interactions between water, vegetation and land use. In an urban context too, physical and human geography are shown to be interlinked when Indigenous scholars and elders teach about dynamic interactions between land use by diverse nations, settlers and immigrants, and the effects of present-day Eurocentric policies. In Chicago for instance, scientific approaches to geomorphology and fluvial systems are integrated with Indigenous realities, so the geography of the city is taught as a palimpsest of Indigenous lands and specific settler colonial policies (Tuck et al. 2014; Bang et al. 2014; **Chapter 4.II**).

[1] For further information on the Dechinta Centre, see https://www.dechinta.ca.

VII Decolonial Understanding and Multi-Epistemic Literacy

In conjunction, the practices described above – of decolonizing the curriculum, working outside the classroom, and confronting racism and dehumanization – seek to reorient student-learners towards global solidarity and familiarity with plural world knowledges (De Lissovoy 2010; Langdon 2013). In this vein, the Portuguese decolonial scholar Boaventura de Sousa Santos (2017) calls for the artisanal construction of an '**ecology of knowledges**', pragmatically adapted to global-local circumstances. According to these principles, decolonial understanding denotes a humble awareness of what we do not know and an impetus to mutual respect for other knowledge-holders. This double move entails paying attention to ethics, and to student-learners' attitudes.

Decolonizing aims to forge more globally attuned, solidarity-based and ethical positionalities towards interpersonal relations and non-western epistemologies (De Lissovoy 2010). Complementing geographies of global responsibility (Massey 2004), decolonial educationalists view global ethics as a counter to sanctioned ignorance. A globalized ethics is shaped through care, responsibility and pragmatism. This transformative ethics, Parvati Raghuram and co-authors suggest, requires 'a more embodied pragmatic responsiveness, one that makes a "care-full" recognition of postcolonial interaction' (Raghuram et al. 2009: 11). In practical terms this means that learning about epistemicide and dehumanization will take place with care for students and teachers and with justice for those affected. Decolonial education moreover recognizes that students come to the classroom with – and express – very diverse experiences and attitudes, in a continuum from privilege to marginalization. In this context, decolonizing seeks to provide students with the tools to navigate these differences respectfully and confront their role in interpersonal oppressions. Listening to a classmate and putting one's own assumptions on hold become valuable skills, providing a route to discussing intersectionality and complex positionalities (Katundu 2020). The aim is for students to think about 'other positions at their own pace rather than defending their own' (Nakata et al. 2012: 134). Nevertheless, such classroom interactions may create sensations of discomfort for learners and instructors alike. As in wider society, decolonizing the classroom brings to the fore strongly held emotional and social reactions that reflect colonial-modern dynamics (Fanon 2008 [1952]). When instructed on colonial-modernity, many students feel uncomfortable and guilty (Kobayashi 1999), feelings that 'manifest in myriad ways including anger, frustration, hostility, antagonism, denial, sorrow and pride' (Daigle and Sundberg 2017: 340). In this context, instructors are responsible for establishing clear boundaries for the expression of emotion and hostility, and working towards intercultural relations.

Achieving decolonial understanding speaks to teaching and learning about complex place-specific geographies of colonial-modernity. Developing frameworks for understanding the spatial, social and epistemic variations in colonial-modernity (**Chapter 1.II, Chapter 3.III**) therefore requires consideration of student progression through a programme. Over two, three or four years of a tertiary-level programme, pedagogies can move from core principles to in-depth analyses across diverse places and groups. **Multi-epistemic literacy** refers to the goal that students – and instructors – be well-versed and equally respectful across a number of epistemologies (that is, frameworks for establishing what we know) (Kuokkanen 2008). Grounded in the most relevant histories and pedagogies for learners and teachers, the goal is an appreciation of an ecology of knowledges. Jettisoning an exclusively western frame of reference, multi-epistemic literacy navigates through transdisciplinary and diasporic comparative approaches (Scott Lewis 2018).

Actions to decolonize the curriculum frequently remain enmeshed in the material relations that sustain the university and public uses of education. The monetization of higher education impacts students, teachers and social relations by cutting across decolonizing imperatives. These limitations occur not only in wealthy countries, as efforts to decolonize and indigenize universities from Ecuador to South Africa face uphill struggles.

VIII Chapter Summary

Calls to decolonize the curriculum and to restructure university education have often prompted interest and engagement in decolonizing debates. As this book argues, decolonizing comprises much more than changing reading lists for students, and this chapter has highlighted the breadth and multifaceted nature of decolonizing education. Decolonizing includes transformations in topics taught, the sources of information on those topics, learner–instructor relations, and expectations about the purpose and outcomes of learning. In this sense, decolonizing the curriculum plays a central part in decolonizing geography, across the discipline from physical to human geographies. With lively debates and the increasing availability of material, the possibility to mainstream decolonizing geography has never been greater. By providing students with information about epistemicide, colonialism and racialization, curriculum decolonization introduces them to 'a pluri-verse of meanings and concepts ... [and] inter-epistemic conversations ... [that] produce new re-definitions' (Grosfoguel 2013: 89).

Nevertheless, 'participation in an educational system that is structurally flawed does not produce decolonization', according to geographer Richie Howitt (2001b: 154). Movements to decolonize the curriculum formed by students and teachers have had a notable and welcome impact on higher level education, including in geography, since the turn of the century. Yet decolonizing the curriculum in geography has not been accomplished; as with other facets of decolonizing, delinking geography from colonial-modern institutional arrangements, social hierarchies and assumptions involves ongoing action and thought. Thus the furthering of decolonizing involves recognizing plural ways of knowing the world, dismantling white privilege, hiring diverse teachers and collating plural forms of content and encounters for students. This may lead tertiary-level geography education into more experimental and solidarity pedagogies and encourage students to engage with diverse knowledges.

Further Reading and Resources

Readings

Bhambra, G.K, Gebrial, D. and Nişancıoğlu, K. (eds) 2018. *Decolonising the University*. London, Pluto.

Cupples, J. and Glynn, K. 2014. Indigenizing and decolonizing higher education in Nicaragua's Atlantic Coast. *Singapore Journal of Tropical Geography* 35(1): 56–71.

Esson, J. 2018. 'The why and the white': racism and curriculum reform in British geography. *Area*, https://doi.org/10.1111/area.12475.

Websites

UCL Collective. 8 reasons the curriculum is white. https://novaramedia.com/2015/03/23/8-reasons-the-curriculum-is-white

Class exercises on whiteness: a resource for instructors to explore with students the impacts of whiteness on everyday life [Based on US context] https://culanth.org/fieldsights/teaching-race-with-lisa-anderson-levy-intersectionality-paradigm-shifts-and-the-ubiquity-of-whiteness

What I learned in class today: Aboriginal issues in the classroom https://intheclass.arts.ubc.ca
This twenty-minute video documents Canadian Indigenous teachers' and students' experiences of devaluation, violence and stereotyping in the classroom. The website also has learning resources including student role-play exercises, information on the project, full interview records, and discussion topics (social position, tokenization, classroom incidents, thinking strategies).

Podcast

Decolonizing the curriculum in theory and practice
https://sms.cam.ac.uk/collection/2345401
University of Cambridge Centre for Research in Arts, Social Science and Humanities (CRASSH), 2016–17. A series of six multidisciplinary episodes addressing the questions: Why is the demand for decolonization being heard so widely in universities today? What place does decolonizing the curriculum have in broader demands for decolonizing the university? What are the experiences of decolonizing the university curriculum in different parts of the world? What would it mean to decolonize the curriculum in Cambridge?

–6–
Decolonizing Geographical Research Practice

Research, the Māori educationalist Linda Tuhiwai Smith notes in her book *Decolonizing Methodologies*, is a 'dirty word' for Indigenous and other colonized peoples (Tuhiwai Smith 2012). As marginalized and subaltern groups across the world know, colonial-modern research is extractive, asking for information that is later used to misrepresent or further damage them (de Sousa Santos 2014). Reflecting on her critique, Tuhiwai Smith argues that in the long run the point is to remake research in less colonizing ways by co-designing it with disenfranchised groups and nuanced **consent** (Tuhiwai Smith 2020). The discipline of geography contributed to making data gathering and research into disempowering and materially damaging processes, because it has treated the world as an arena for imperial expeditions and a site for collecting data for its own purposes, all the time sanctioning ignorance of its negative implications. Decolonizing geographical research prompts delinking from such practices and forging relationships of responsibility and listening. Decolonizing does not necessarily invalidate *or* prioritize 'overseas' or cross-cultural fieldwork; rather, wherever it occurs, research needs to affirmatively counter racialized dehumanization, colonial structures of governance and polarizing political economy (Daigle and Ramírez 2019).

In this sense, decolonizing geographical research implies decentring canonical processes in favour of less familiar and more open-ended collaborations with people, among them groups and individuals marginalized in previous knowledge production. With these objectives, decolonizing transforms the entire research cycle, from the formulation of initial ideas to choosing methodologies, addressing ethics, fieldwork and writing up and sharing findings. In the long haul of decolonizing, personal and institutional commitments are needed to 'reframe, reclaim and rename' (Louis 2007: 132) the reasons for research, its circulation and audiences, and the agency and engagement of participants. This chapter primarily addresses geographers who undertake research in the context of a university, who are racialized as white, and/ or are located in a wealthy country, as it is 'our' practice that needs to change most urgently in student projects, PhDs and professional research. The chapter discusses this type of geographer and research in general, with specific pointers for student projects and doctoral research. The situations of researchers of colour, Southern researchers and community-based researchers raise different issues that are discussed in key texts (e.g. Denzin and Lincoln 2014).

Gathering and interpreting new information was an integral part of colonial control, being driven by the 'heuristic and documentary requirements of a metropolitan and administrative leadership' (Simpson 2007: 67). Colonial data collection continues into the colonial present as 'a powerful remembered history' (Tuhiwai Smith 2012: 1). Physical and human geographies retain colonial understandings of the value of western, scientific and detached enquiry. The basic categories geographers use in their research are often – if mostly unintentionally – deeply colonizing. Globally endorsed theoretical and conceptual frameworks underpin the expectations and interactions with research participants or populations in the 'field', making their experiences and understandings of marginal or no interest. Black and Indigenous groups and people of colour have to make their knowledges recognizable on northern normative terms, adjustments that ultimately reproduce epistemic violence and foreclose opportunities for empowerment. Too often, the powerful undertake research 'on' the less powerful (human geography) or 'on' their lands and environments (physical geography). These people and places end up 'ghettoized, orientalised ... and ... overstudied' (Tuck and Yang 2014: 223), which determines the type of information collected and its audiences (Mutua and Swadener 2004: 12). Around the world, neoliberal university agendas and countries' geopolitical ambitions exacerbate and entrench these dynamics (Noxolo 2017a).

Yet despite its complicity with power, research is important as it remains 'one of the last places for legitimated inquiry ... [and] cares about curiosity' (Tuck and Yang 2014: 223). Linda Tuhiwai Smith (2020) has not stopped doing research; instead she constructs Indigenous research capacity and trains Indigenous graduates, working with community-run research institutes in Aotearoa-New Zealand. Geographers take up the challenge to rethink disciplinary traditions of research, as examples in this chapter show. Steps to decolonize research build on decades of critical geographies and speaking truth to power, enriched by feminist, postcolonial and participatory approaches, among others (de Leeuw and Hunt 2018). Recently, the momentum towards decolonizing has extended these efforts in a more systematic rethinking of research away from privileged positionalities. Physical and human geography's 'view from nowhere' impoverishes understanding, so decolonizing involves building relationships with the peoples and places where we do research, and recognizing coloniality in physical landscapes and socio-spatial relations. On the basis of respectful relationships, dialogue about the scope, purpose and nature of research can begin.

This chapter presents reasons for and practical steps towards decolonizing geographical research, with examples from human and physical geography. It is organized around the life-cycle of a research project, from foundational principles and goals through to preparing, organizing and defining data collecting, then writing and sharing research. Although these stages are found in standard research, decolonizing fundamentally reworks the goals, decision-making processes and practices of research from start to finish. Boxes provide in-depth discussions of undergraduate research projects, graduate student fieldwork and ethical protocols for work groups and individuals under colonial-modernity.

I Decolonizing Research: Principles and Goals

Decolonizing is – as noted throughout this book – a long-term process of challenging the mindsets and material realities of coloniality, and has no single endpoint. Hence decolonizing research means challenging colonizing practice, re-devising principles to generate new knowledge-producing practices, and taking responsibility for undoing colonial-modern geographies (Sundberg 2014; Daigle and Ramírez 2019). Fundamentally, this means switching from a mode of extraction to a thoughtful two-way process between all participants (Louis 2007: 135). Indigenous, feminist and decolonial geographers argue that decolonial research engages with the structural and historically endorsed ways knowledge has been made, circulated and validated. Whether for a student doing their first project (**Box 6.1**), a PhD fieldwork session, or a tenured academic, the challenge involves building relationships with research participants in new ways, based on acknowledging one's privilege, and beginning with flexibility, accountability and communication to ensure processes that minimize harm and re-humanize research.

With decolonizing, the non-academic groups and individuals involved in a project become participants, interlocutors and partners (*not* 'subjects' or 'informants'), as their concerns, worldviews and analyses inform the purposes of the research. Physical geographers focus on landforms, ecosystems and fluvial systems (among other things), approaching them as objects for quantitative scientific methods. Yet these same features are meaningful places for diverse local social groups, each with engagements that differ from those of physical geographers (except for the few non-white, non-northern physical geographers). Decolonizing encourages us to pause and recognize the systems of colonialism and imperialism that led to physical geography's disavowal of social positionality and its dismissal of non-western types of understanding. For instance, geographers who study glaciers rarely 'recognize indigenous knowledges, local perspectives, or alternative narratives of glaciers, even though large populations of non-Western and indigenous peoples inhabit mountain and cold regions near glaciers and possess important knowledge about cryoscapes' (Carey et al. 2016: 773). Decolonizing, by contrast, engages physical geographers with parallel knowledges to their own, and rethinks methods and analysis in that light.

Human geographers are often just as likely to ignore or belittle alternative knowledges even when human relations and socio-environmental relations are the focus. In human geography, the 'blinkers' of white and geopolitical privilege are unsettled when researchers (metaphorically or literally) 'walk with' other knowledge-holders (Sundberg 2015: 123). In physical or human geography, these steps may involve learning non-dominant languages and distinct world-knowledges. Otherwise, the challenge is to uncover the subtle and pervasive power relations that sideline certain experiences, world-knowledges and interpretive frameworks. In each case, 'translation' between epistemologies occurs through uneven institutional, epistemic and social relations of power. Mutual esteem, clear accountability and respectful representation become the cornerstones of decolonizing relationships based on listening, face-to-face experiences and continual dialogues (MacDonald 2017; Barker and Pickerill 2019). This section provides an introductory overview of these decolonizing processes.

Orlando Fals-Borda's work in Colombia in the late 1960s provides a powerful starting point for decolonizing research. Finding positivist methods counterproductive for understanding Latin American inequalities, Fals-Borda advocated the following pointers, as applicable in human as in physical geography:

• Do not monopolize your knowledge or impose your techniques arrogantly but respect and combine your

skills with the knowledge of the researched or grass-roots communities, addressing them as full partners and co-researchers
- Do not trust elitist versions of history and science which respond to dominant interests, but be respectful of counter-narratives and try to do justice to them
- Do not depend solely on your culture to interpret facts, but engage with local values, ways of living, beliefs, and arts for action
- Do not impose a scientific style for communicating results, but share what you have learned together, in a manner that is wholly understandable and even literary and pleasant, for science should not be necessarily a mystery nor a monopoly of experts and intellectuals (adapted from Munck 2020: 137)

Fals-Borda's principles remind us forcefully that decolonizing is as much about acknowledging the structural factors that enable individuals to undertake research as it is about seeking out alternative knowledge (whether with people, landscapes or ecosystems). Researchers 'occupy specific temporal and sociocultural positions, positions often bound to or by colonialism' (de Leeuw and Hunt 2018: 3). As discussed in previous chapters, geographers' knowledges of the world reflect colonial-modern geopolitics and racial hierarchies, meaning their partial knowledge and ignorance are sanctioned and become normalized (Spivak 1999; **Chapter 3.I**). The geographers Daigle and Ramírez (2019) call for geographical research that affirmatively refuses white supremacy, anti-Blackness, settler colonial states and racialized political economy. The UK-based human geographers Jazeel and McFarlane (2010) suggest that geographers adopt an open, learning positionality to undo Eurocentric university-centred claims to expertise. This is an uncomfortable process, as western (especially white) geographers gain from the institutional endorsement of their 'expertise' (Castleden et al. 2017). Decolonial research replaces this with humility and listening, participant oversight and control, and decolonial thinking from start to finish. In this respect,

Box 6.1 Short student projects: learning to decolonize

Doing a short research project with decolonial objectives in mind does not restrict the topic, scope or questions to be addressed. A decolonizing strategy involves critical reflection on the project's rationale and your personal positioning in place-specific epistemologies and colonial-modernity. Each piece of research exists within particular financial, academic and logistical limitations; decolonizing asks us to broaden the scope of design and preparation. Addressing directly how your project emerges from colonial forms of geographical thinking and the urgency of decolonizing clarifies the planning and implementation processes.

Questions to ask yourself in advance: It may not be immediately clear or self-evident where and how coloniality-modernity plays out in a place or for a specific group. Identifying this takes persistence and willingness. Ask yourself about the reasons for the research:

- How is coloniality involved in your position and in relation to the topic you are interested in?
- Is it possible to fulfil your university's requirements and do something that helps others?
- Could your research empower disadvantaged individuals or groups (how could you check with them)?
- Have you considered 'do no harm' and accountability in ethical planning (**section 6.IV** on ethics)?
- Is it possible to liaise with an organization requesting a student-sized project? (Many university professors are already involved in collaborative research, while some student associations and outreach programmes have links to local civil society.)
- How far in advance should you start talking with research partners, and agree on the scope and outcomes?

- Are there alternatives to working with 'overstudied' areas and groups? For example, could you turn to (un)official archives, existing databases or digital sources (see Ferretti 2019)?
- How will you share your research results with partners (e.g. brochures; data and archives left with partners; digital stories)?
- What is your responsibility to participants, alongside academic progress and university requirements?

Undertaking to decolonize a short research project brings (manageable) challenges, including limited time for planning and data gathering, having less established peer-to-peer relationships with non-university groups, and being obliged to work within your institution's procedures. Support from instructors, shared reading lists and open discussion about strategies can mitigate these challenges.

decolonizing processes are as applicable to physical as to human geography.

Experimentation

Decentring from western complacency and working with marginal groups and landscapes, however, does not lead to a handy check-list outlining the next steps. Due to the diverse forms of colonial and imperial violence, oppressions and forced changes in knowledge experienced across the colonial-modern world (**Chapter 1.II**), decolonial projects have to carefully work *with* this variegation and adapt to context and history (**Chapter 3**). This calls for decolonizing experimentation, not in a scientific testing sense, but in interaction with partners based on *their* priorities, knowledges and protocols in order to devise a mutually relevant and more open-ended research methodology. Decolonial knowledge creation thus embarks on a new form of 'experimentation, boundary-crossing and risk; it is likely to be interdisciplinary, sometimes radically so' (Connell 2014: 215). This will often result in an unprecedented and unique approach. Its experimental modality articulates with the specific circumstances of the research (not universal procedures) by supplementing and/or reframing existing approaches.

Two examples illustrate how decolonizing is concerned with changing researcher–partner relationships. Based in the Americas, the Other Knowledges Transnational Network RETOS defends social movements as creators of knowledge, using methodologies and analysis that reject the western subject–object divide underpinning colonial research in favour of relational, plural and critical approaches (Maldonado-Torres 2016; RETOS 2018). On the other side of the Pacific, Australian geographers and Aboriginal partners collaborated in 'open-ended social and ecological experiments' to better understand how distinct bushfire knowledges could be brought into conversation. With Aboriginal partners retaining control over information, this experimental approach had 'results and effects which cannot be fully known in advance' (Neale et al. 2019: 344). Decolonial research works pragmatically across plural ways of knowing, each way knowing better or more about itself than others, but coordinating to address a specific question (Castleden et al. 2017). For example, biodiversity science, peasant and Indigenous knowledges pool their insights to better understand how to protect biodiversity (Baker et al. 2019; Turner 2020).

Experimentation relies on trust and communication. On the one hand, research questions and analyses are not derived solely from university-based theory, as decolonizing is open to new priorities and epistemologies that challenge western models and expertise. Research becomes committed to place-contextualized agendas and meanings, thereby breaking colonial patterns of expropriation of peoples' and environments' knowledges (Nhemachena et al. 2016). Experimentation also disrupts university geographers' position as coordinators and managers, while teams of academics, community participants and leaders multiply the range of commitments and roles, always requiring flexibility in local implementation and university-funders' processes (Tipa et al. 2009). Shifting to undo colonial violence, the decentring of university researchers' roles represents a reordering of knowledge production (Smithers Graeme and Mandawe 2017).

Decolonizing knowledge creation requires each and every mode of knowing to become experimental, in a collaboration based on deep listening, pragmatic flexibility and analytical adaptations.

Humanizing research and building accountability

Working across social difference has a long history in human geography, whether with north Atlantic scholars carrying out fieldwork in developing countries, or urban and social geographers working in varied class, racial-ethnic and sexual communities. Unspoken colonial legacies infuse many research programmes, which are approached through the normative lenses of whiteness, colonial-modern hierarchies and Anglophone universities' authority. The decolonial philosopher Maldonado-Torres (2016) blames these dynamics on colonial knowledge relations between researchers (western, often white, often male), the object (being studied), and methods of distanced observation. They result in dissociation from and denial of full status to those being researched (in some cases, denial of parity to places and environments). Denying full humanity and dignity to less powerful individuals, groups and places results in their negative characterization and enhances western self-identity (Valentine 2002). At their most blatant, colonial relations of knowledge construct Africa as a site for gathering raw data, and Africans as medical or social scientific 'guinea-pigs' (Nhemachena et al. 2016).

In these circumstances, colonial knowledge relations generate what the African-American scholar bell hooks terms 'pain narratives' (hooks 1990) that portray the marginalized as agency-less victims. Indigenous and decolonial commentators argue that these narratives oversimplify and depoliticize realities, to the detriment of those affected (Louis

2007). In material terms, the search for such narratives results in an overstudying of certain groups, whose under-resourced leaders experience fatigue and frustration as *their* concerns remain unaddressed (Tuck and Yang 2014; LaRocco et al. 2019). Decolonizing research hence entails the systematic acknowledgement and disruption of colonial knowledge production and white supremacy. However, decolonizing alternatives *are* available and feasible. One PhD student found she 'couldn't bear to perpetuate colonial demands on the exhausted community leaders' (Coddington 2017: 316), so switched to alternative sources and archival methods. Another possibility is 'studying up', that is, researching the powerful groups, individuals and institutions that reproduce and gain from coloniality. Here the researcher's authority as western, university-based and often whiter facilitates access to respondents and data inaccessible to others (Zahara 2016).

Decolonizing in this sense means delinking from geography's status quo, and finding ways to recognize coloniality while seeking to dismantle it in thinking and practice. Replacing anti-Indigenous, anti-Black re-colonizing practices (Asselin and Basile 2018) with relationships forged with people and places whose participation is key to the research shifts definitively away from an extractive, subject-object-method mode. People and places at the site of research become interlocutors and agents in producing, holding and transmitting knowledge. As discussed below (**section 6.II**), this approach provides more nuanced and disaggregated information and bases for analysis. Indigenous geographers argue that building attachment and proximity to territory and land works to delink from dehumanizing universals and aids learning with alternative epistemologies (Simpson 2007; Daigle and Ramírez 2019).

Respecting the integrity of different knowledge systems in a research site is key in decolonizing research, as it seeks to build bridges in order to establish interactions and insights that break out of a one-world world geography (Johnson et al. 2016). Building bridges for decolonial outcomes can take different forms, from community involvement in designing interview questions, to participants' involvement in analysis and writing up the research (**section 6.V**). Alternatively, a decolonial project may create a report or outcome that reflects diverse community participants' priorities and goals as well as university research (Zurba et al. 2018). For example, a doctoral student with longstanding, protocol-led relations with a South American Indigenous group co-authored with leaders a summary account of their historical, political and organizational trajectory. The report provided the group with a training and organizational resource, and the student with a validated account to inform the PhD.

As researchers, we inhabit particular (colonial-modern) positions for which we must take responsibility (Sundberg 2014). To humanize research, decolonizing makes explicit the obligations to participants, thereby making researchers 'responsible, not to a removed discipline (or institution) but rather to those studied' (Denzin and Lincoln 2014: 19; CDRE 2013). In decolonizing, there are no universal contexts and experiences, so research must engage in a systematic and sustained learning about the place-specific protocols that will guide researcher–participant relations. In some cases, these protocols are formalized rules or socially meaningful practices and expectations (**section 6.IV**). Addressing responsibility and accountability in-depth strengthens group-specific and national protocols on human and more-than-human thriving (Daigle and Ramírez 2019; de Leeuw and Hunt 2018; Bawaka Country et al. 2020; **Table 6.1**). Overall, delinking from colonial research must become mainstream across the entire discipline, in both physical and human geography.

Principles of decolonizing research: in summary

Decolonizing agendas thus guide researchers to humanize and build accountability, which provides the basis for respectful, plural and ethical research practices. Although decolonial research is location-specific, Indigenous scholars and critical geographers highlight a set of core principles to be adhered to across all research. These principles work towards relationships and procedures for critical relational accountability,

Table 6.1 What makes decolonizing research?
The table compares key features of research and decolonizing approaches in geography.

	Standard geographical research	Decolonizing research
Relation with marginalized knowledges and groups	Through Eurocentric representations and western social theory	Respectful representation, nuanced understanding of multiple colonial-modernities
Researcher–researched relation	Viewed as informants, who will transmit data to researcher	Viewed as collaborators, involved in reciprocal relations
Positionality in knowledge systems	University and professionals have expertise in research	Researcher engages in unlearning then re-learning through engagement with other knowledges
Analysis and sharing of findings	Eurocentric analysis, research distributed via academic journals and meetings	Collaborative analysis, peer review by participants, distribution in appropriate ways to participants, negotiated community ownership of data
Collaborative ethics to challenge white privilege and Eurocentric criteria of knowledge	Racialized institutional structures and interactions are perpetuated	Collaborative process among participants informs a contextualized ethics informed by anti-racism and a decentred geopolitics of knowledge

Source: Adapted from Louis 2007; Simpson 2007; de Leeuw and Hunt 2018; Vaeau and Trundle 2020.

justice-orientation and openness to plural experiences in knowledge production:

- Respectful of diverse epistemologies
- Relevant to participants
- Reciprocal
- Responsible

Additionally, decolonizing debates suggest further principles for research practice:

- Rights (partners' rights have highest priority)
- Sustained horizontal **Relationships**
- Returning (results and in person)
- Ongoing critical **Reflection** (Snow 2018)

Together, these principles provide guiding reminders of the priorities of decolonial research, without prescribing how or with whom that research is undertaken (Riddell et al. 2017; CIHR, NSERC and SSHRC 2018).

II Decolonizing Research Design

Decolonial research retains a vivid awareness of the historically entrenched power dynamics and plural epistemic context of a project's theme and locations. Researchers are frequently enmeshed in westernizing universities, work in fieldwork sites scarred by colonial-modernity, and are expected to distribute results internationally. For the geographer Juanita Sundberg, this means that there is 'no disinterested place' from which to do research, as it starts from 'our own sites of entanglement and engagement' (Sundberg 2015: 123). Decolonizing research thus aims at open-ended collaborative principles, as outlined above, yet is continually confronted in practice by the tensioned research context. This section examines strategies for dealing constructively with these tensions, looking at decolonizing rationales, rethinking participatory research, understanding research **refusal** and becoming a scholar ally.

Decolonizing research rationales from the start

Decolonizing research begins at the very start of thinking about a project, rather than 'adding in' elements at the stage of data collection or analysis (Riddell et al. 2017). As discussed in Chapter 3, geography is well-placed to analyse and critique colonial-modern spatial relations. Accordingly, a decolonial research project needs to understand a specific place and situation, building on questions such as:

Box 6.2 The novice researcher and decolonizing processes

PhD and Masters students are frequently novice researchers with few or no personal histories of working in a research project. In comparison with established scholars they may lack social networks with partners. Graduate researchers hence face specific hurdles when adopting and completing a decolonizing project. Thesis committees and supervisors may be uninformed or sceptical about decolonizing approaches, leaving a student without constructive engagement. Strategies to address this situation include a department-wide debate about decolonial geographers, using meetings with advisors to speak about coloniality in research, and/ or looking 'outside' to allies (without over-burdening them). University review boards often misunderstand the need to gain participants' ethical approval before and then again after official ethical review. Finding allies among tenured staff who have secured flexibility from the university on this point assists graduates. Graduate projects face particularly strong pressures on resources and time, with scholarships and university regulations setting the limits on open-ended decolonial discussions. The often large distances between a graduate's university and their fieldwork site make it difficult to sustain face-to-face conversations at settings and times appropriate for participants. Overall, decolonizing graduate research works to delink from westernizing expectations and arrangements that disempower partners, and to shift from the production of a data-rich thesis towards an appreciation of plural epistemologies and accountability to knowledge-holders beyond the university.

Sources: Castleden et al. 2012a; MacDonald 2017; Snow 2018.

- Where is the coloniality in this context?
- How might this topic gain from pluralized knowledge, working especially with border knowledges?
- Which research questions are most relevant to the dispossessed, silenced and racialized individuals and groups in this context?

These concerns prevent geographers looking at the world as if it were merely a set of potential field sites (Sundberg 2015) (**Box 6.3**). Biogeographers have begun important conversations about how relations with field sites need to consider responsibilities and researcher–local relations (Baker et al. 2019; Bannister 2020).

Box 6.3 Decolonizing physical geography: water catchment research

A research group comprising Māori and non-Māori geographers examined 'stream health' in Aotearoa-New Zealand, to track the cultural and biological status of the water catchment (Tipa et al. 2009). On a Māori initiative, a place-specific and culturally appropriate research design was agreed by all participants. It emerged as a result of numerous conversations and was eventually written into a formal partnership statement. The agreement specified the methods for data gathering, chosen to be appropriate to the various epistemologies involved, yet the overall project placed Indigenous priorities front and centre. The project occurred on parallel tracks – university and Māori – with individual and collective activities ranging from literature reviews to journeys to meaningful places. Information was gathered in aural and written forms, aerial photos and the embodied experience of site visits. On completion, the team agreed to report the findings through diverse outlets.

Due to the colonial-modern structure of university geography, understanding how coloniality has influenced – and continues to shape – places and situations involves seeking out information from sources and perspectives that appear at first sight remote from the topic. This 'homework' (Sundberg 2015) is integral to decolonial research preparation. Doing homework prompts critical self-reflection on one's presumptions in order to avoid taken-for-granted frameworks and reinforcing power inequalities. Homework informs the researcher about colonial-modern contextual realities 'to enable ethical decisions about what to research, with whom, using which practices, and to what ends' (Sundberg 2015: 122; see also Blunt and Wills 2000: 181–7). Homework may additionally prompt reflections on whose knowledges could inform the research design (de Leeuw and Hunt 2018). In other words, homework initiates an open-ended, experimental and decentred process that adds critical awareness of colonial realities to the geographers' conceptual, analytical and methodological decision-making processes.

Participatory research and decolonizing agendas

Many geographers adopt participatory modes of research as it permits a closer and often collaborative relationship with interlocutors. Community-based participatory research (CBPR) is widely perceived as an ethical mode of engagement, as it shares decision-making power between researchers and participants. However, complicated power dynamics crosscut participatory research, as has been noted by African and Indigenous geographers. Decolonial and Indigenous critiques draw attention to its limitations, and seek to address wider critical challenges associated with knowledge production. In standard forms, participatory work invites community or individual participation only after the researcher has designed the project around university or disciplinary-defined goals. As a result, it puts individual participation at risk when the main priority is maintaining group knowledge and information (Bawaka Country et al. 2020), or uses participation to access a 'community' without considering the (heterogeneous) group's information needs. In a set of Canadian CBPR projects with First Nations, non-Indigenous researchers used decolonizing participatory principles of respect, relevance, reciprocity and responsibility more consistently during the phases of data collection and sharing findings than during the project design and data-analysis stages (Castleden et al. 2012a).

In this and other so-called participatory research, 'participation or action in research does not amount to liberation' (Nhemachena et al. 2016: 23). For this reason, scholars are increasingly clear that participatory collaboration is not sufficient as decolonial practice. Decolonizing participation involves the inclusion of participants in early shared decision-making about the questions, methods, purposes and outcomes of research. It means university-based geographers – each with epistemic baggage and material privilege – acknowledging how decolonizing transforms and complicates interactions between collaborators (Coddington 2017; Meer and Müller 2021). Integral to decolonizing is a systematic and thorough transparency about the basis for participation, so that interlocutors have control over its nature and possible consequences. Extensive dialogue in a considered decision-making process, consulting on the allocation of tasks and responsibilities, ensures a fuller understanding of the whole project on all sides. Decolonizing thereby reorders researcher–community relations:

> recognition of participants as researchers and pedagogues with agency even as they participate in informal researcher/teaching roles asks us to re-imagine research as a non-hierarchal teaching/learning/advocacy process rather than a method of investigation and discovery which echoes violent colonizing projects of history. (Gill et al. 2012: 11)

To fulfil decolonizing potential, participation needs to be woven *throughout* a project in order to build sustained control and partners' oversight.

Refusal

Despite best intentions, the colonial realities of research alienate and disengage misrepresented groups and individuals, whose situation is worsened because of it (Simpson 2007; Vaeau and Trundle 2020). Communities can be unwilling to participate, particularly with university-based academics, for reasons related to control over research processes. Distrust is often widespread, arising from diverse facets of colonial power, and may lead to a refusal to take part in research. Such refusal can be expressed verbally as well as through boredom, frustration, silence, avoidance, evasion, and non-attendance at pre-arranged meetings (LaRocco et al. 2019). Refusing extractive research sets limits and asserts control over the making and sharing of knowledge, thereby refocusing geographers' attention on the multifaceted epistemic and structural violence wreaked in the name of research. Refusal empowers the groups affected by extractive research, as it asserts their autonomy over their knowledge and marks it as off limits (Simpson 2007; Tuck and Yang 2014: 223).

For these reasons, refusal is taken seriously in decolonizing scholarship, requiring the researcher to ask themselves: What is being refused and why? What needs to change? (Tuck and Yang 2014: 21–2). The Kahnawake Mohawk anthropologist Audra Simpson argues that refusal can generate innovative action by redirecting researchers' focus onto critiquing colonial structures. In some cases, this leads to greater accountability through joint decision-making (in geography, see Zahara 2016; Coddington 2017). 'All refusal is particular' (Tuck and Yang 2014: 243), as it reflects time- and place-specific histories and experiences. Non-Māori researchers in Aotearoa-New Zealand are enjoined to 'not try to "overcome" the "barrier" of mistrust but rather sit firmly within it and let it lead the research towards an appropriate approach' (Vaeau and Trundle 2020: 209). Due to refusal's particularities, responses will depend on context. Taking the possibility of distrust and refusal seriously at the research-design stage alerts researchers to the factors leading to refusal, and suggests routes for resolution. Negotiation and dialogue aid in overcoming distrust, while incorporating and affirming participants' agency to delink research from colonialism (Sylvestre et al. 2018).

Becoming a decolonizing scholar ally

As these discussions highlight, undertaking research is a far from neutral, liberal intervention in peoples' lives and places' ways of being. Human and physical geographers have to reflect on their interactions, perhaps deciding to become allies of groups and landscapes (water systems, glaciers, ecosystems …) that face active colonialism. Marginalized and overstudied groups are increasingly asserting their rights to define and organize data gathering and analysis for themselves. These moves prescribe how, when and where scholars and students can undertake research work. Indigenous organizations in North and South America, for instance, frequently require written agreements, especially for university and cross-cultural collaborations. Such arrangements seek to overturn extractive research and instead establish co-working and transparency about data control and circulation. University-based geographers may seek to support these initiatives, becoming allies by endorsing critical epistemologies, valuing autoethnography and insider method-ologies, and 'construct[ing] stories in the landscapes through which we travel' (Denzin and Lincoln 2014: 8). For white geographers, becoming a **decolonial ally** involves critical self-reflection on racialized power. Overall, such alliances aim to forge 'praxis and inquiry that are emancipatory and empowering' for those with least power (Denzin and Lincoln 2014: 2), while the decision to trust a scholar ally rests with subaltern actors (Asselin and Basile 2018; Snow 2018).

This section has focused on pragmatic responses to the inherent contradictions of attempting decolonial research in a world of colonial-modern knowledge production. Decolonizing research asks the researcher to position themselves in relation to coloniality-modernity and prompts thinking beyond existing participatory approaches. The historical legacies of colonial-modern research and its present-day consequences can give rise to refusal. Nevertheless, because of these tensions, decolonial research remains relevant by inaugurating dialogues with partners.

III Designing Methodologies

Methodology refers to the principles and assumptions underlying the choice of methods for constructing and analysing data (Gregory et al. 2009: 457). Methodology is about finding the appropriate methods (techniques for generating information) to address research questions, in light of the overall objectives of the research. This section outlines the criteria for designing a decolonial methodology consistent with decolonizing epistemology, i.e. with how we know the (colonial-modern, 'one-world') world and what we should know (plural perspectives). Designing a methodology occurs before the choice of methods and involves decisions about how the information to be generated will contribute to decolonizing (Johnson et al. 2016).

In light of coloniality's pervasive influence, decolonizing encourages methodologies that provide sensitive yet robust data-collection techniques and enable wider socio-economic-political benefits. 'Decolonizing methodologies bring together, among others, Indigenous, transformative, liberation, feminist, and [critical methodologies] to strengthen decolonizing research' (Brendon Barnes quoted in Leibowitz et al. 2019: 28). Decolonial methodologies are coherent with the decolonizing principles (**section 6.I**). They seek to change the relationships through which information is garnered and shared, which in turn transforms geographical knowledge production (de Leeuw and Hunt 2018). Methods such as interviews remain important in decolonizing research, although their implementation, context and purposes change. There is no single 'off the shelf' methodological toolkit, as each project needs to create a contextualized, process-appropriate and transparently discussed methodology.

These issues have been extensively discussed by Indigenous scholars, although today the term 'Indigenous methodologies' has become shorthand for co-constructing methodologies with any group or set of individuals who experience racialization, dehumanization, colonial-modernity and de-legitimation (Denzin and Lincoln 2014). Indigenous methodologies may include Indigenous and other practices already used for gathering and systematizing information for their purposes and, as such, comprise context-appropriate tools, especially as they reflect and endorse Indigenous knowledge systems (see Tuhiwai Smith 2012 on twenty-five Indigenous research projects). For this reason, acknowledgement of Indigenous, African-American, Pacific Islander and Latinx methodologies is a 'critical step in decolonizing geography' (Louis 2007: 136). Researchers and interlocutors co-construct decolonizing methodologies through dialogue around the most appropriate research questions and ways to generate knowledge (Asselin and Basile 2018).

Indigenous methodologies provide alternative ways to think about research, but do not reject all mainstream approaches (Louis 2007: 133). They may contribute to a project that includes parallel tracks of community data gathering and academic geographical methods. Alternatively, a set of methods may be adopted for goals shared by the researchers and participants (Tipa et al. 2009). Deciding on methodologies often involves an ongoing dialogue between participants and researchers, starting with trust-building and open conversations. For a research proposal I wrote after conversations with national Indigenous women leaders in Ecuador, I suggested that Kichwa and Tsáchila women record daily audio diaries. However, local leaders rejected this approach and suggested instead that the women do face-to-face interviews with me during break-outs at community meetings. As I came to understand, this methodology was premised on socially embedded knowledges (accessed via my observation of community meetings) and did not burden the already overworked women. By getting me to listen to debates (conducted in local idioms and contexts) in collectively constituted spaces, the research became a means to record women's relational, politicized and group-constituted information, and immersed me in their epistemologies.

In a Aotearoa-New Zealand context, Māori principles of *whanaungatanga* (building and maintaining responsible, reciprocal and committed relationships) and *manaaki-tanga* (showing care and respect for a two-way flow of knowledge) are proposed for anti-colonial and decolonizing research (Vaeau and Trundle 2020). Whichever method-ology is adopted, research partners own and control the data-gathering process, confirming their central involvement throughout the research cycle (Denzin and Lincoln 2014). Methodologies such as starting-point stories (Riddell et al. 2017) go further in ensuring participants define in a collective and informed way how and with whom information will be gathered, further decentring the researcher's control.

A range of methods including photo voice, autoeth-nography, visual methods, storytelling, community-based participatory action research, archival work, critical personal narratives and interviewee-verification are apt ways to recon-figure power and encourage respectful sharing in decolonial research. These methods centre experiences and interpreta-tions and make space for narratives and critiques which are less accessible or 'audible' in standard methodologies (Denzin and Lincoln 2014; MacDonald 2017). In order to overturn misrepresentations, decolonizing methodologies consider in depth how the capacity to elicit and nurture infor-mation is highly dependent on the spaces in which methods are deployed. Participant-led encounters with places and with more-than-human actors can play a significant role in shaping the quality and nature of data gathering, and are relevant for both physical and human geography. For this reason, participant-led 'walking' and site-visit methods are used to indicate their trajectories and priorities (**Table 6.2**).

Decolonial geography seeks to broaden and restructure the power relations inherent in research, as well as become better able to examine and understand processes and experi-ences largely hidden from university-based Anglophone geography. Decolonizing methodologies are deliberately flexible as they must be suitable for the site of research and for documenting variegated and exclusionary colonial-modern processes.

Table 6.2 Decolonizing research methods

This indicative list is far from exhaustive, as a variety of place-relevant data-collection methods are used and advocated in decolonizing geographical research (see Tuhiwai Smith 2012).

Method	Description and applicability in decolonizing research	Further information and examples of uses
Autoethnography – critical personal narratives	Centres postcolonial/subaltern researchers' accounts of how they come to know, name and interpret experiences in the world. Reclaims subjectivity and voice, especially of hybrid, ambiguous and in-between experiences and processes.	Chawla and Atay 2018 Ellis, Adams and Bochner 2011
Photo voice	Working with participant-led narratives, participants use a digital camera and a context-appropriate research guide to document their narratives. Later, participant selection and interpretation of photos occurs within their expertise and experiences.	Mark and Boulton 2017
Storytelling	A participant-led and participant-created story-based data-gathering strategy, to address the limitations of interview-based narratives. Multimedia digital storytelling illustrates personal narratives and stories with photographs, artwork, music, voice-overlay, video clips and text. Stories are created during immersive workshops. The development and sharing of personal stories celebrates individual and collective experiences.	Meer and Müller 2021; Cunsolo Willox et al. 2012; Wright et al. 2012
Interviewee verification (repeated interviews)	Over a series of interviews, the interviewee is asked each time to verify the researcher's understanding and interpretation of previous conversations. Interviewees control the information exchange. Repeat interviews build relationships and trust.	Snow 2018; Radcliffe 2015
Drawings	Participant-led description and interpretation, engaging their priorities and agency, situated in place and material circumstances.	Leibowitz et al. 2019
Body-territory mapping	On a drawn outline of a (gendered) body, participant-led mapping of experiences and territory-wide processes that begin and end in bodies. This generates participant-led discussion of the implications of capitalism, patriarchy, racism and colonialism, viewed through an intersectional lens.	Vela-Almeida et al. 2020; Zaragocín and Caretta 2020
Participant-led walking interview	Research participants lead the researcher around their chosen places and sites. Walking interviews elicit embodied and situated knowledges and information.	Warren 2021

Source: Author

IV Ethical Issues and Dilemmas

Geographical associations and university departments have thorough and thoughtful statements and procedures to ensure professional ethics.[1] The discipline has welcomed in-depth discussions about the norms, ethics and politics of research procedure, in which colonial-science-type frameworks are thoroughly critiqued. Nevertheless, decolonizing is about more than improving ethical standards (Barker and Pickerill 2019), as universities and the geography discipline continue to support research that underpins colonial-modern power. Geographers rightly place great emphasis on ethical review processes. Learning about research ethics and applying ethical principles to a specific project comprise a central plank of graduate education, and often of undergraduate geography programmes. As decolonial geography stresses researchers' responsibility and accountability to participants, so too the 'ethical space' (Gentelet et al. 2018) of project planning engages new dimensions. Three aspects of this dynamic are covered here: negotiating ethical issues between research participants and universities; dealing with consent; and research participants' oversight of a project's ethics. The discussion hence moves from university-based issues, towards researcher responsibilities, and finally to participants' authority over ethical questions.

University ethical review processes reflect a particular type of taken-for-granted approach, using a 'generic, utilitarian, biomedical, western model of ethical inquiry' (Denzin and Lincoln 2014: 19). In this context, ethics procedures primarily respond to authorized and (Eurocentric) scientific knowledge production in the institution. Ethical approval thus tends to

[1] For example, the American Association of Geographers has a statement of professional ethics at http://www.aag.org/cs/about_aag/governance/statement_of_professional_ethics. The New Zealand Geographical Society's code of conduct specifies an additional duty of care to people to ensure participants are not disadvantaged in any way. Research in New Zealand additionally has to accord with the Treaty of Waitangi. See https://www.nzgs.co.nz/code-of-conduct.

protect a more metropolitan and disciplinary-based domain, and power relations that grant the researcher decision-making privileges (CDRE 2013). Moreover, standard ethical review boards are not routinely attuned to ask questions about how a project addresses structural issues of racialization and coloniality (Coddington 2017). By contrast, a decolonizing project involving co-labour with groups outside the university may incorporate social and environmental ethical principles and procedures. In this situation, a decolonial project would have to undergo an autonomous and participant-driven set of ethical checks and approvals as well as a university review (e.g. Snow 2018). Despite profound differences between university and partners' procedures however, decolonial projects negotiate between them, for instance by undertaking repeated reviews by partners and the university, and then by partners again.

Decolonizing research grapples with reworking the *purpose* of ethical reviews and the *conversations* through which projects get approval. Whereas the university sees a procedure, the researchers and partners experience an ongoing dialogue about power, control and outcomes. In geographical research, gaining consent from an interviewee or survey respondent is the cornerstone of ethical practice, to ensure their understanding of the research's use of responses. Standard ethical guidance usually seeks a respondent's consent to use individually held knowledge; an individual signature is accepted as sufficient. By comparison, decolonizing engages with participants whose sense of knowledge, power and responsibilities do not necessarily correspond to this framework. Among groups with knowledges subjected to epistemicide and extractive research, information-sharing is not decided by a single person, or without prior group consideration of access. Hence, decolonial research often involves repeated conversations and interviews for which consent is required each time. Participant-led projects with Indigenous and other groups increasingly create a negotiated and context-specific process to arbitrate over who can share which knowledge, with whom and for what purposes (**Box 6.4**). In one project, conducting research on bushfire

Box 6.4 Sample consent form and agreement

The consent form below was used with Indigenous peoples and First Nations in Canada. It constitutes an agreement that the community and researchers must follow in the storing and sharing of confidential information. All partners are accountable to participants for fulfilling the terms of consent, which are accepted in writing or verbally by each individual interlocutor. The researchers were allowed to keep the data for seven years, and communities decide how long to keep their data (with participants' agreement). Based on each participant's consent form, researchers created a dataset for each partner community. A copy of the data was given without identifying information for participants who wished to remain anonymous, and with identification of those consenting to use of their data in future projects (Riddell et al. 2017: 14, 16).

CONSENT FORM

Would you like to participate in the project?   _ gave approval
                                                                          _ declined

Audio-recording:
I agree to be audio-recorded   _ Y _ N
I agree for my audio-recordings (which can identify me) to be included in what is shared with my community   _ Y _ N

Confidentiality (please choose one statement):
_ I agree to be identified by name/be credited in writings and/or presentations OR
_ I prefer not to be identified by name; please use _____ OR
_ I prefer to be identified as 'community member' and a code chosen by the team

Future use of what you share today:
I consent to the use of the knowledge I shared today in future research: _ Y _ N
I consent to be contacted again if the knowledge I shared today is requested for use in future research: _ Y _ N

(Source: Riddell et al. 2017: 16)

management, Neale and Aboriginal co-authors (2019) established protocols to ensure Aboriginal knowledges remained entirely within their domain of control. In the design of decolonial ethical processes, clear rules about what can be collected, held and then analysed and shared are central.

Anonymity for respondents is crucial to protect privacy and, in some cases, personal security. Decolonizing research which acknowledges the ongoing violence of coloniality on groups and places needs to exercise especial caution to protect participants' lives (Hale and Stephen 2013). However, where research is oriented towards recognizing and circulating otherwise-silenced voices, participants may seek public recognition of their contributions, and wish to be named. This occurred with one of my Kichwa co-researchers after numerous conversations about anonymity. Ultimately, the issue of anonymity must be negotiated, with information-sharing about the context for research dissemination.

These issues speak to the centring of participants' priorities and rights. While university review boards address some of these issues, participant groups seek further negotiation and control over processes and findings. Indigenous groups, particularly in the Americas and Australasia, require formal agreements or oversight committees comprising community members, leaders and non-academic allies (Wright et al. 2012). In each case, the process remains in the hands of research participants who choose and ratify agreed measures. Following national or community-led frameworks (Table 6.3), memorandums and agreements lay out the exact processes to be followed in designing the project, data collection, contributions from community members and leaders, data analysis and sharing findings (see e.g. Ioris et al. 2019; Castleden et al. 2012b). To lessen the risk of epistemic violence, agreements honour participants' categories and knowledge, and acknowledge them as complex and dynamic (Bawaka Country et al. 2020). These protocols aim to avoid any depoliticizing technical application and co-optation in the name of knowledge 'co-production' (Sundberg 2015; Johnson 2018).

Table: 6.3 Codes of best practice: Indigenous and decolonial research

Sustained engagements with Indigenous people in collaborative research practice are currently found in Canada (with national Indigenous organizations including the Assembly of First Nations, the Inuit Tapiirit Kanatami, the Metis National Council, the Native Women's Association of Canada and various urban Indigenous organizations, for example the Ontario Federation of Indigenous Friendship Centres). The table cannot cover all areas so it summarizes recent policy and guidelines with decolonizing features. As best practice is continuously refined by Indigenous, Black and peoples of colour, researchers are encouraged to identify protocols and materials where they do research.

	Year	Organization	Highlights
Ownership, Control, Access and Possession (OCAP®)	2005	First Nations Information Governance Centre (https://fnigc.ca)	Indigenous approach arising from Indigenous research and activism. Community control of research process and data; community choice to establish a partnership; emphasizes cultural and social importance of knowledge; community focused.
Tri-Council Policy Statement: Ethical Conduct for Research Involving Humans (TCPS2)	2010 (updated in 2014, 2018)	Canadian Institutes of Health Research; Natural Science and Engineering Research Council; Social Sciences and Humanities Research Council	Guidelines written with Indigenous communities. Research should enhance capacity of Indigenous groups to maintain cultures, languages and identities; acknowledges the intentional/unintentional harm done by research; mandatory compliance; nature of community engagement must be described in university ethical checks; must respect Indigenous governing bodies. Further discussion needed for partnership approach.
Urban Aboriginal Knowledge Network framework	2012	National Association of Friendship Centres, and Urban Aboriginal Knowledge Network	Utility with relevant and direct benefit; self-voicing affirms communities as authors and knowledge-holders; access recognizes that all forms of knowledge are reliable and valid; inter-relationality notes that research always occurs in historical and relational contexts. Ongoing negotiations on consent; ensure fairness, respect and honesty. Indigenous intellectual property rights, and profit returned to Indigenous partners.
Data sovereignty issues	McAlvay et al. 2021	Native Nations Institute	Indigenous governance database; sovereignty through various open-source software and web-based platforms.
Charter of Decolonial Research Ethics (CDRE)	2013	European social movements	Right to suspend research at any time and to review all writing before sharing; explicit permission required for publication.

Source: Adapted by the author from Riddell et al. 2017; Castleden et al. 2012a: 164; CDRE 2013; McAlvay et al. 2021.

One feature in geographical research is a reliance on gatekeepers and community-based research associates (RAs). Especially in cross-cultural work, where the researcher is not a member of a participating group, these individuals play a key role in legitimating and facilitating the social relations that underlie all successful research partnerships. RAs facilitate trust-building (especially important in research collaborations with overstudied groups), act as 'translators' across knowledge systems, and ensure participants' oversight. However, the researcher–RA–gatekeeper relationship is mired in colonial-modern dynamics, especially when a western university researcher relies on individuals with rich knowledge and networks without providing appropriate employment status or co-authorship. These relations moreover are often crosscut by postcolonial intersectional relations of gender, race, national origin, class and level of formal education. In this context, negotiated and agreed terms are necessary for inclusive, respectful relationships (Smithers Graeme and Mandawe 2017). Raising and discussing these points before the project commences contributes to transparency, accountability and social justice (LaRocco et al. 2019). Formal agreements hardwire in the priorities and expectations of research associates, including co-authorship.

Decolonizing ethics are firmly embedded in wider principles of respect for and responsibility to participants and to lands and territories. While sometimes this leads to tensions with university reviews or to dual/parallel ethical review processes, decolonial research priorities remain those of context-specific accountability and transparency and, as appropriate, formal agreements and employment contracts.

V Analysis, Writing and Sharing

Analysing data, writing up the findings and then sharing results with audiences are central in producing and validating geography's knowledges, whether in human or physical geography. Standard – colonial-modern – geographical practice, however, has not considered how research participants can and should be systematically included in these stages of geographical research.

Decolonial and humanizing arguments highlight the crucial element of providing research partners with first access to project findings and with control over distribution (Denzin and Lincoln 2014). This might mean involving local civil society and relevant organizations in physical geography and biogeography projects, and local populations and associations in environmental and human geography data analysis and distribution. Participant involvement in analysis makes interpretation more attuned to the epistemic and socio-political context, which in turn triangulates research and makes it more robust. However, sufficient time is required for participants to learn about, reflect on and provide feedback on initial findings. At the planning stage, calculating the required time is crucial. Participant analysis of findings adapts to context, and need not replicate standard disciplinary analytical approaches. For instance, initial findings can be presented for discussion at workshops and focus groups, as well as in written or visual forms. The decolonial goal is to create an inclusive space and opportunity for generating feedback and contributing new points that will shape an emerging analysis. Invitations to groups outside the university signal respect for non-university epistemologies (MacDonald 2017). Decolonizing analysis may also involve using informants' words and concepts (with permission) as the basis for coding, and may involve iterative interactions to refine those codes and interpretations. Likewise, repeated interviews with the same group give respondents the opportunity to correct researchers' initial interpretations (Snow 2018: 7). Another decolonial strategy is to co-author longer pieces to record findings together, which puts data analysis, interpretation and presentation more firmly in the hands of research partners (de Leeuw and Hunt 2018: 3).

The process of sharing results requires careful consideration. In one Aotearoa-New Zealand project, researchers agreed in advance on diverse means to inform publics, including posters and brochures. Academic publications came last, taking into account the different priorities of Māori

Box 6.5 Aymara Indigenous control over writing and sharing

The sociologist Silvia Rivera Cusicanqui has long worked in and outside Bolivian universities and alongside racialized and dispossessed Indigenous populations to generate knowledges which are generally excluded from national histories and school curricula (**Chapter 2.III**). She and researchers at THOA (the Andean Oral History Workshop) required everyone to speak Aymara before collecting rural Indigenous personal narratives. Participating Aymara groups had the right to decide on whether and how to allow THOA's findings to be disseminated. They chose to broadcast a radio soap opera in Aymara, as radio was a major media source for rural low-income groups.[1]

[1] For an English-language interview on THOA's context and activities, see http://www.oralhistoryforsocialchange.org/blog/2019/6/6/oral-history-can-bring-us-into-a-longer-arc-of-resistance-interview-with-benjamin-dangl.

and university participants (Tipa et al. 2009). Following decolonial best practice, prior discussions with participants clarify priorities for sharing and communicating results, and decide on authorship of various outputs (e.g. RAs, communities and/or Aboriginal Country where research occurred) (LaRocco et al. 2019). Consistent with decolonizing is the choice to submit material for publication only after participants grant explicit and informed permission and following peer review of material (CDRE 2013). Although time-consuming, participant review can identify misinterpretations, thereby strengthening the research, although ultimately responsibility for writing – especially when done for student qualifications – lies with the researcher. Although academic publications in western-oriented journals do not overcome colonial-modern knowledge production (Connell 2007), they can respond to the priorities and longer-term goals of project partners, for example by providing leverage in lobbying and public information campaigns.

Another aspect of decolonial writing is its acknowledgement of contributors' positionalities and where they speak from. This is especially helpful in indicating the range and type of collaboration across diverse epistemologies. A glossary of local terms contributes to communication across epistemic difference, making the material more accessible and distributable. Overall, decolonizing processes continue to the last stages of a research project, sustaining a respectful and accountable dynamic with participants.

VI Chapter Summary

In relation to research, it is useful to remember that decolonizing is an ongoing process and not a single endpoint. Colonizing research unthinkingly adjusts itself to dominant priorities and paradigms while its knowledge production occurs in self-referential frames, especially under neoliberal pressures. Research often starts with a set of goals, yet in practice makes continual adjustments, takes unexpected turns and involves learning. In this sense, both research and decolonizing are profoundly iterative processes that require continuous reflection on our practices and assumptions, and time to pause and adjust processes and relationships. Overall, as this chapter has shown, there is no template for a decolonial research project, as each will involve multiple actors and diverse forms of knowledge, and will be embedded in uneven, place-specific landscapes of power. Nevertheless, geographers around the world and in different career paths have a number of tried-and-tested tactics in working towards decolonizing research. Learning from these resources makes for better research and better researchers, enabling the production of critical reflective knowledge.

Decolonizing research asks us to slow down and rethink the design of research projects, their collaborative nature and the collective ethical environment in which they happen. It invites us to work together to dismantle white privilege and

Eurocentric notions of knowledge. These steps do not remove the need for flexibility and taking responsibility; rather, they re-situate the implications of research for participants and communities. Decolonizing thus entails broadening and pluralizing dialogues and ethical considerations in a process that is *of* the colonial-modern world yet retains the capacity to challenge it. An epistemic space is being 'carved out' in decolonizing geographical research, although on the ground individual projects reflect a highly differentiated colonial-modern world and diverse knowledge-production agendas. In the words of Bawaka Country and co-authors, 'we do not, cannot, have answers that are applicable in all contexts. We acknowledge, and ask that you acknowledge, the laws, protocols, positionalities of your place' (2020: 998).

Further Reading and Resources

Readings

Brown, L. and Strega, S. (eds) 2015. *Research as Resistance: Revisiting Critical, Indigenous, and Anti-Oppressive Approaches*. 2nd edition. Toronto, Canadian Scholars' Press. (This collection provides foundations in specific methodologies, and draws out their anti-oppressive and emancipatory potential, for final year undergraduates and starter postgraduates. Taking location as a central concern, chapters cover queer, Indigenous, feminist, critical race theory and community action research.)

CIHR, NSERC and SSHRC. 2018. *Tri-Council Policy Statement: Ethical Conduct for Research Involving Humans*. Ottawa, Institute of Health Research.

Elder, B. and Odoyo, K. 2018. Multiple methodologies: using community-based participatory research and decolonizing methodologies in Kenya. *International Journal of Qualitative Studies* 31(4): 293–311.

Woodward, E. and McTaggart, P.M. 2016. Transforming cross-cultural water research through trust, participation and place. *Geographical Research* 54(2): 129–42.

Websites

Decolonising methodologies, 20 years on
https://www.thesociologicalreview.com/decolonising-methodologies-20-years-on-the-sociological-review-annual-lecture-by-professor-linda-tuhiwai-smith
The Sociological Review Annual Lecture by Linda Tuhiwai Smith, 16 October 2019, Goldsmiths, University of London. The website provides context and discussion of the lecture, which can be viewed at https://www.youtube.com/watch?v=YSX_4FnqXwQ.

Decolonizing the interview method
https://decolonialdialogue.wordpress.com/research
R. Ghemmour, Decolonial Dialogue blog

Urban Aboriginal Knowledge Network
http://uakn.org/wp-content/uploads/2014/10/Guiding-Ethical-Principles_Final_2015_10_22.pdf
Guiding Ethical Principles, UAKN-RCAU, 2015

Decolonizing research under Covid-19

African Academy of Sciences. 2020. *Research and Development Goals for COVID-19 in Africa: Priority Setting Exercise*. Available at www.aasciences.africa.
Corbera, E., Anguelovski, I., Honey-Rosés, J. and Ruiz-Mallén, I. 2020. Academia in the time of COVID-19: towards an ethics of care. *Planning Theory and Practice* 21(20): 191–9.

Glossary

The glossary provides brief definitions of the terms associated with decolonizing geographies and interdisciplinary decolonial thinking. Given the rapidly changing discussions around decolonizing, the vocabulary and definitions provided here necessarily reflect their time and place. Readers may find the terms and definitions do not match those in their part of the discipline or world, and additional terms and concepts may occur to them. The pluralizing and open-ended processes of decolonizing make this an inevitable and welcome outcome, in a world where many worlds fit. Some entries use terms that appear elsewhere in the glossary; these are highlighted in bold.

Anglophone geography – The academic discipline of geography in English-speaking countries and territories.

anti-colonialism – A series of political ideas, organizations and actions opposed to **colonialism** and **neocolonialism**, which gained global reach and influence from the nineteenth century.

anti-Indigenous racism – Overt and covert forms of misrepresentations of Indigenous and First Nations peoples that reproduce symbolic and institutional violence.

anti-racism – A series of political ideas, organizations and actions opposed to diverse forms and facets of **racialization**.

autonomy – A high level of decision-making power and control over society and its setting, particularly in relation to **Indigenous peoples'** political transformations in nation-states. Autonomy can take a number of forms, from legal provisions in **settler colonial states** through to radical strategies for **self-determination**. Autonomy is not equivalent to or a step towards **sovereignty**.

BAME or BME – Acronym for Black, Asian and Minority Ethnic populations, a term used widely in UK media, government and education.

BIPOC – Acronym for Black, Indigenous and People of Colour used in North American media and social activism.

Black geographies – A body of research and theory developed by and for Black peoples' territories, places and spaces.

border knowledges/border thinking – Coined by Gloria Anzaldúa, the concept refers to thinking from both inside and outside the **colonial-modern** system. It arises from a different geopolitical and bodily positionality to Eurocentric **universalism**.

canon – An established body of knowledge which implicitly or explicitly sets the standards against which new knowledge is judged.

capitalism – A form of economy and social organization with a concentrated ownership of resources among one class, and wage labour for the majority. Capitalism as an historic and globalizing economic system became the focus of analysis from the late eighteenth century, coinciding with **imperialism** and **colonialism**. See also **racial capitalism**.

colonial matrix of power – Originating with Aníbal Quijano, the term identifies facets of **coloniality-modernity**. The matrix consists of **coloniality of power** (inherent racial hierarchies in global capitalism), knowledge systems of the **one-world world**, and coloniality of being (subjective experiences of global inequalities).

colonialism – The control of people and territory by another country, which often uses self-justifying ideologies such as civilization or racism. Colonialism extracts resources and wealth through land occupation, cultural-identity norms and political-legal measures, sometimes granting **settlers** territorial and social control.

coloniality – The mindsets, knowledges, identities and

structures of power that originate in **colonialism** and interlock with modernity. Coloniality comprises 'longstanding patterns of power that emerged as a result of colonialism but that define culture, labour, intersubjective relations and knowledge production well beyond the strict limits of colonial administrations' (Maldonado-Torres 2007: 243).

coloniality-modernity – Aníbal Quijano's concept developed by Walter Mignolo. It draws attention to the inseparability of modernity and **coloniality**, in which modern ways of thinking (**epistemes**) contain a logic of coloniality which justifies violence against non-western, oppressed and **racialized** populations, knowledges and alternative power.

coloniality of gender – The organization of gender in the modern-colonial world. Decolonial feminists suggest that coloniality restructures gender in relation to race and facets of identity and embodiment through violence.

coloniality of power – The sociologist Aníbal Quijano's concept of a global **hegemonic** model of power, in place since the conquest of the Americas, which articulates race and labour, space and peoples, according to the needs of capital and the benefit of white peoples.

consent – Indigenous peoples assert control over territories and resources yet lack power in **settler colonial** and **postcolonial** societies. International law and Indigenous campaigns seek Indigenous decision-making rights over projects, development and interventions. This principle involves free (un-coerced), prior (before intervention occurs), informed (relevant information is available to affected groups) consent (interventions only occur if and when Indigenous groups agree).

cosmopolitanism – A desire to think about all human beings as the basis for moral concern. Eurocentric cosmopolitanism is premised on racial hierarchies, excluding groups considered as less than human. Non-western notions of cosmopolitanism include Buddhism, African notions of Ubuntu, and **MCD** alternatives.

critical race theory – An established body of knowledge and theory, originating in North America, that offers a framework for understanding racism. Critical race theory (acronym CRT) suggests **racism** is embedded in social structures, has a material foundation, changes and develops over time, and comprises more than individual prejudice.

curriculum – The lessons and academic content taught in an educational institution.

decolonial ally – A geographer who works collaboratively with emancipatory and empowering practices based on critical and decolonial ethics and objectives.

decolonial turn – A set of critical theoretical and political interrogations of power emerging in the early twenty-first century. The philosopher Nelson Maldonado-Torres distinguishes it from previous anti-colonial and decolonial expressions, describing it as 'an increasingly self-conscious and coalitional effort to understand decolonization, and not simply modernity, as an unfinished project' (2011: 2).

decoloniality – Thinking and acting to undo, disobey and delink from **coloniality** and seek alternative modernities. In geography, decoloniality refers to understanding and acting to change the spatial differentiation and socio-spatial relations that arise from coloniality.

decolonization – The struggles of populations and leaders to gain political independence as **sovereign** states from colonial masters through highly variable means.

decolonizing – A process of active delinking from **coloniality**; 'a long-term process involving the bureaucratic, cultural, linguistic and psychological divesting of colonial power' (Tuhiwai Smith 2010: 33).

deep colonizing – A form of power characteristic of educational institutions that appears committed to reversing **coloniality**, but which in practice reproduces it.

dehumanization – The denial of full humanity and status (legal, social, epistemic). **MCD** scholar Maldonado-Torres (2007) argues that dehumanization arises in **coloniality** that violently links notions of racial difference to institutions, practices, interpersonal relations and knowledge. Decolonial feminists argue that dehumanization is structured through gender and racial difference.

doctrine of discovery (*or* discovery doctrine) – European legal principles used to justify and legitimate the annexation and colonization of non-European (specifically non-Christian) lands and peoples. The discovery doctrine has been used in the Spanish and Portuguese colonization of the Americas, and diverse forms of **settler colonialism**.

ecology of knowledge – The legal scholar Boaventura de Sousa Santos's term for a methodology to bring plural knowledges into dialogue in order to challenge **colonial-modern** hierarchies between knowledge systems.

enslavement – Practices making a slave of a person or group, depriving them of political freedom, often associated with exploitation of their labour.

episteme – A principled system of understanding, and a tacit framework for theory and empirical work (**epistemology**). Decolonial scholars argue that **hegemonic** colonial-modern epistemologies do not preclude the existence of **border knowledges,** in contrast to Foucault's view that only one episteme defines the conditions for knowledge at any one time.

epistemic disobedience – A deliberate break from dominant Western-Eurocentric thinking by engaging with plural, 'marginal' knowledge systems.

epistemic violence – The postcolonial theorist Gayatri Spivak's concept refers to the violence done to non-western ways of knowing and understanding throughout colonialism and beyond.

epistemicide – Boaventura de Sousa Santos's term for the destruction of non-western knowledges and worldviews resulting from Eurocentric and colonial actions, categories, logics and ways of knowing about the world.

epistemology – The study of how we know the world and how it ought to be known; a dominant epistemology creates the framework to legitimate and reproduce certain kinds of knowledge at the expense of others.

Eurocentrism – The complex formations of knowledge and science that reflect the European region's historical experience and which became globally **hegemonic** after seventeenth-century colonialism through **coloniality-modernity.**

geographies of decoloniality – A multifaceted theoretical framework that seeks to understand and explain the spatial differentiation and socio-spatial relations that arise from **coloniality** and decolonizing actions and thought (see **decoloniality**).

geographies of racialization – Geographical analysis of how processes of **racialization** constitute and rationalize spatial exclusions, relations and patterns on the basis of **racism** and associated hierarchies.

geopolitics of knowledge production – The term refers to the power of hegemonic countries to determine how knowledge is gathered, formulated and validated. **MCD** scholars highlight the social and institutional positioning of knowledge systems in relation to geopolitics and **intersectionality**.

global South – A majority of the world's countries in Africa, Latin America, Asia and the Caribbean which, despite disparities in income, politics and society, are contrasted to wealthy 'northern' countries. The term originated in the 1980 Brandt Report, and marked the transition from Cold War and 'Third World' non-alignment geopolitics.

hegemony – The process of exercising power and control through the spread and acceptance of dominant groups' and institutions' values. **Colonialism** resulted in the hegemony of **Eurocentric** forms of governance, social orders and worldviews.

imperialism (empire) – A geopolitical relation where a state controls the political **sovereignty** of another state, through formal (including military) or informal (cultural, economic) means (Slater 2004). Imperialism is a broad category including US post-war influence, direct colonization and diverse forms of **colonialism**.

indigeneity – The qualities associated with being Indigenous, including both **colonial-modern** assumptions attributed to **Indigenous people** (tradition, timeless attachment to land, etc.) and politically strategic characteristics mobilized in Indigenous movements.

Indigenous geographies – A body of research, praxis and theory developed by and with Indigenous peoples in relation to their territories, places and spaces.

Indigenous peoples – The United Nations uses the term to refer to peoples' self-identification as populations with longstanding connections to lands, socio-political-cultural distinctions, and agendas of **self-determination**. The labels 'Indian' and 'Aboriginal' originated in European colonialism with negative connotations. Indigenous peoples have created international networks for rights and seek to nurture diverse languages, ways of living and **epistemologies**.

intersectionality – A concept developed by Kimberlé Crenshaw and feminists of colour to identify and critique interlocking exclusions arising from being women and Black. Intersectional theories draw attention to **relational**, qualitative, substantive and hierarchical powers of race, gender, sexuality, class, location and other differences.

knowledge – see **epistemology, Eurocentrism, knowledge production**

knowledge production – Decolonial approaches identify hierarchies between knowledge systems which are fundamental for coloniality's power, as western institutions, presumptions and scientific and ethnographic practices predominate over those of others. Decolonizing knowledge production involves structural change in institutions, assumptions and practices for respectful dialogues between knowledge systems and **multi-epistemic literacy**.

microaggression – Non-white racialized groups' lived, everyday experiences of subtle, often unconscious, nonverbal and behavioural indignities, causing visceral and psychic impacts.

modernity-coloniality – see **coloniality-modernity**

modernity-coloniality-decoloniality (MCD) – A diverse group of scholars and activists, largely from Latin America and the Caribbean, who focus on understanding and explaining the persistence of coloniality in modernity, and interrogating colonial-modern production of knowledge about the world (see **geopolitics of knowledge production** and **knowledge production**).

more-than-human – An Anglophone geography term for the intricate, dynamic interconnections between geo/Earth, bio/life and human society. Indigenous and other scholars and activists criticize the term's reliance on western science and technology studies that **sanction ignorance** – and the scholarly appropriation – of Indigenous ontologies in which the more-than-human is central to life-worlds.

more-than-one-world – A term devised for this book to indicate the scope of decolonial geographies. Informed by critical geographical and decolonial approaches, the term signals that geography's **decolonial turn** reorients the discipline towards multiple geographies, a world in which many worlds fit, and actively challenges the **one-world world**.

multi-epistemic literacy – The decolonial goal that students, and instructors, become well-versed in and equally respectful towards various and diverse **epistemologies**.

neocolonialism – A term used by Ghana's first president to refer to western powers' continued domination of post-independence African countries. Latin American dependency theory developed similar critiques.

neoliberalism – Economic theory and political governance premised on market-led growth and reductions in state investment and welfare. Influential globally since the early 1970s, neoliberalism is implicated in maintaining and extending colonial labour and resource relations.

Occidentalism – A term coined by the Venezuelan anthropologist Fernando Coronil for European forms of classification, hierarchy, exclusion, naturalization and spatiality, deployed in colonial global power.

one-world world – The historian of science John Law discusses how European colonialism collected information about the world, evaluated it on European criteria, and decided it was universally valuable. The one-world world doctrine represents a self-referential system of understanding the world, using only western methods and approaches (see **epistemicide**).

ontology – 'The study and description of "being" or that which can be said to exist in the world' (Gregory et al. 2009: 511). Decolonial scholars suggest that Eurocentric ontologies are exclusionary and unrepresentative of the world, and advocate learning about and from other ontologies (see **colonial matrix of power**).

Orientalism – The Euro-American construction of the Middle East through texts and images, argued by Edward Said (1978) to be integral to colonial power relations. Orientalism generated imaginative geographies, practices and institutional arrangements 'in advance of and in conjunction with colonialism, underwriting colonial power' (Gregory et al. 2009: 513–14).

Other – A critical term identifying a group or person perceived as fundamentally different to a western, 'modern', metropolitan person or group. Otherness reflects power and justifies exclusion, denying social connections and context-dependent identities. An other ('Other') is often located spatially.

Othering – Processes leading to binary representations of (social, racial, sexual and so on) difference, based on unequal power and a denial of interconnections.

Pachamama – Andean Indigenous concept referring to a complex living **more-than-human** entity comprising the earth, physical features in landscapes, animals and plants as well as human and other-than-human beings.

paradigm shift – A significant transformation in methods, values and intellectual assumptions shared by a group of scholars (or scholar-activists) that shifts the focus and content of **knowledge production**.

pluriverse – A totality of possible worlds and multiple ways of making life and being. The opposite of western **universalism**, the concept refers to a dynamic intermingled multiplicity of world-making. Hence the adjective pluriversal.

postcolonial – Denotes the contexts defined by colonialism, viewed from a critical perspective.

postcolonial condition – The situations of life in the aftermath of colonialism and colonization, acknowledging the mutually constitutive role played by colonizers and colonized populations. The postcolonial condition pervades political, economic, social, spiritual, cultural and knowledge relations.

postcolonialism – Critical scholarly approach from the 1980s criticizing and contextualizing colonial discourses and representations. Geographical postcolonialism analyses how space is represented, organized and imagined in the colonial-imperial past and postcolonial present.

racial capitalism – The world-spanning socio-economic system in which **capitalism** operates through **enslavement**, genocide, appropriation and criminalization. **Racism** and capitalism are integrally connected, resulting in **dehumanization** and exploitation.

racialization – Historically and spatially variable processes by which racial power and meanings are established and maintained in social, political, economic, psychological and epistemic relations. Frantz Fanon identified the socio-psychic effects of the colonial racialization of Blackness.

racism – Overt and covert relations of power premised on social understandings of biological differences between human groups. Racism results in material and epistemic hierarchies, and is routine in colonial-modern interactions and institutions.

refusal – Assertion of control by misrepresented groups to prevent damaging and 'extractive' research.

relational – Critical social science term for the primacy of relations in the making of beings (society, landscape, nature ...).

sanctioned ignorance – A concept developed by Gayatri Spivak to refer to how colonial assumptions define the western **canon** through criteria that close down consideration of other forms of thought.

self-determination – In international law, the principle that a people should decide how they will be governed. Since mid-twentieth-century decolonization, groups – primarily Indigenous peoples – have claimed rights to self-determination. International law (e.g. ILO Convention 169) seeks to define forms of self-determination without undermining existing nation-states' **sovereignty**. Self-determination is also associated with **autonomous** government in sub-national territories, e.g. Nunavut, Canada.

settlers – Incoming groups or descendants of immigrants in

a previously occupied territory, who reside permanently in that territory and take action that dispossesses prior residents.

settler colonialism – A form of colonialism whereby **settlers** (primarily but not exclusively Europeans and their descendants) move deliberately to a territory in order to remain (settle) there and to displace existing – including **Indigenous** – populations. Critical scholars and activists argue that settler colonialism seeks to eliminate prior residents through replacement and/or assimilation.

Southern theory – The sociologist Raewyn Connell's (2007) term for a range of thinkers and social theories in formerly colonized societies who rework western theory in light of regional realities. Southern theory, Connell argues, is rarely taken into account by **Eurocentric** scholars and theories.

sovereignty – The authority claimed by states over territories and peoples, giving states the right to pass laws and be recognized by other countries. **Imperialism** and **neocolonialism** shape expressions of sovereignty. Internally, state authority may be opposed and resisted, generating overlaps of state and non-state authorities.

subaltern – Individuals and groups without **hegemony**, whose lives and knowledges are not socially and politically influential. Indian and Latin American scholar-activists have developed understandings of complex colonial and **postcolonial** power relations and the silencing of marginal groups. In decolonial approaches, subaltern describes social groups with lives and knowledges that exist against and within **coloniality-modernity**.

terra nullius – An eighteenth-century European legal doctrine that made unoccupied land or occupied 'unproductive' land available for colonization.

universalism – The idea that certain characteristics, concepts, morals and knowledge are valid for all places and times. Decolonial approaches identify European universalism with **coloniality-modernity**, which suppressed alternative life-ways and knowledge systems.

unmarked power – Power that is not expressed through direct

and visible channels, and instead is ingrained in everyday actions and assumptions. Unmarked power becomes a social norm and sometimes appears 'natural'. Geographers identify unmarked power in **intersectional** relations; anti-racist geographers argue that **whiteness** has unmarked power in the discipline.

westernizing (or westernized) university – A research and teaching institution where **knowledge production** is embedded in Eurocentric **epistemologies** which are treated as objective, disembodied and universal.

white geographies – 'Ways of seeing, understanding and interrogating the world … based on [hierarchical,] racialized and colonial assumptions that are unremarked, normalized and perpetuated' (Domosh 2015: 1).

white privilege – A set of facilitating advantages for people who are **racialized** as white. White privilege influences mobility, authority, opportunities, security, bodily integrity, health and legal standing.

whiteness – A racialized attribute that signifies, embodies and materializes power relations in coloniality-modernity. Frantz Fanon and bell hooks originated the critical study of whiteness. Geographers draw on multiple disciplines to examine whiteness and the place of white bodies in systems of racialized oppression.

Bibliography

Abbott, Dina. 2006. Disrupting the 'whiteness' of fieldwork in geography. *Singapore Journal of Tropical Geography* 27: 326–41.

Abu-Lughod, Janet. 1989. *Before European Hegemony: The World System AD 1250–1350*. Oxford, Oxford University Press.

Adas, Michael. 2016. Colonialism and Science. In H. Selin (ed.) *Encyclopedia of the History of Science, Technology and Medicine in non-Western Cultures*. New York, Springer, pp. 604–9.

Ahmed, Sara. 2000. *Strange Encounters: Embodied Others in Post-Coloniality*. London, Routledge.

Ahmed, Sara. 2007. A phenomenology of whiteness. *Feminist Theory* 8(2): 149–68.

Ahmed, Sara. 2012. *On Being Included*. Durham, NC, Duke University Press.

Al-Saleh, Danya and Noterman, E. 2020. Organizing for collective feminist killjoy geographies in a US university. *Gender, Place & Culture* 28(4): 453–74.

Anderson, Kay. 2008. 'Race' in post-universalist perspective. *cultural geographies* 15: 155–71.

Anthias, Penelope. 2017. *Ch'ixi* landscapes: indigeneity and capitalism in the Bolivian Chaco. *Geoforum* 82: 268–75.

Antipode Editorial Collective (ed.). 2019. *Keywords in Radical Geography: Antipode at 50*. London, Wiley.

Anzaldúa, Gloria. 1987. *Borderlands/La Frontera: the new mestiza*. San Francisco, Aunt Lute Books.

Asher, Kiran. 2013. Latin American decolonial thought: or making the subaltern speak. *Geography Compass* 7(12): 832–42.

Asselin, Hugo and Basile, S. 2018. Concrete ways to decolonize research. *ACME* 17(3): 643–50.

Auerbach, Jess. 2018. What a New University in Africa is Doing to Decolonise Social Sciences. In S. de Jong et al. (eds) *Decolonization and Feminisms in Global Teaching and Development*. London, Routledge, pp. 107–10.

Awâsis, Sâkihitowin. 2020. Anishinaabe time: temporalities and impact assessment in pipeline reviews. *Journal of Political Ecology* 27(1): 830–52.

Baker, Kate, Eichhorn, M.P. and Griffiths, M. 2019. Decolonizing field ecology. *Biotropica* 51: 288–92.

Bang, Megan, Curley, L., Kessel, A., Marin, A. et al. 2014. Muskrat theories, tobacco in the streets, and living Chicago as Indigenous land. *Environmental Education Research* 29(1): 37–55.

Banivanua Mar, Tracey and Edmonds, P. (eds). 2010. *Making Settler Colonial Space: Perspectives on Race, Place and Identity*. New York, Palgrave Macmillan.

Bannister, Kelly. 2018. From ethical codes to ethics as praxis: an invitation. *Ethnobiology Letters* 9(1): 13–26.

Bannister, Kelly. 2020. Right Relationships: Legal and Ethical Context for Indigenous Peoples' Land Rights and Responsibilities. In N.J. Tucker (ed.) *Plants, People, and Places*. Montreal, McGill-Queen's University Press, pp. 254–68.

Barker, Adam J. and Pickerill, J. 2019. Doings with the land and sea: decolonising geographies, indigeneity and enacting place-agency. *Progress in Human Geography* 44(4): 640–62.

Barnett, Clive. 2020. On the abolition of the geography department. Blog post, 1 July. At https://poptheory.org/2020/07/01/on-the-abolition-of-the-geography-department.

Barrera de la Torre, Gerónimo. 2018. Las 'otras' geografías

en América Latina: alternativas desde los paisajes del pueblo Chatino. *Íconos* 61: 33–50.

Bartmes, Natalie and Shukla, S. 2020. Re-envisioning land-based pedagogies as a transformative third space. *Diaspora, Indigenous, and Minority Education* 14(3): 146–61.

Bawaka Country, Wright, S., Suchet-Pearson, S. et al. 2016a. Co-becoming Bawaka: toward a relational understanding of place/space. *Progress in Human Geography* 40(4): 455–75.

Bawaka Country, Burarrwanga, L., Ganambarr, R. et al. 2016b. Co-becoming Time/s: Time/s-as-telling-as-time/s. In J. Thorpe et al. (eds) *Methodological Challenges in Nature-Culture and Environmental History Research.* London, Routledge.

Bawaka Country, Suchet-Pearson, S., Wright, S. et al. 2020. Bunbum ga dhä-yuṯagum: to make it right again, to remake. *Social & Cultural Geography* 21(7): 985–1001.

Bhabha, Homi. 1994. *The Location of Culture.* London, Routledge.

Bhabha, Homi. 2008. Foreword. Remembering Fanon: Self, Psyche and the Colonial Condition. In F. Fanon, *Black Skin, White Masks.* London, Pluto, pp. xxi–xxxvii.

Bhambra, Gurminder. 2014. Postcolonial and decolonial dialogues. *Postcolonial Studies* 17(2): 115–21.

Bhambra, Gurminder, Gebrial, D. and Nişancıoğlu, K. (eds). 2018. *Decolonising the University.* London, Pluto.

Blaser, Mario. 2014. Ontology and indigeneity: on the political ontology of heterogeneous assemblages. *cultural geographies* 21(1): 49–58.

Blaser, Mario and de la Cadena, M. 2018. Pluriverse: Proposals for a World of Many Worlds. In M. de la Cadena and M. Blaser (eds) *A World of Many Worlds.* Durham, NC, Duke University Press, pp. 1–22.

Blaut, J.M. 1993. *The Colonizer's Model of the World.* New York, Guildford Press.

Bledsoe, Adam and Wright, W.J. 2019. The pluralities of Black geographies. *Antipode* 51(2): 419–37.

Blomley, Nicholas. 2003. Law, property and the geography of violence: the frontier, the survey and the grid. *Annals of the Association of American Geographers* 93(1): 121–41.

Blomley, Nicholas. 2006. Uncritical critical geography? *Progress in Human Geography* 30(1): 87–94.

Blunt, Alison and Wills, J. 2000. *Dissident Geographies*. Harlow, Pearson Education.

Bonds, Anne and Inwood, J. 2016. Beyond white privilege: geographies of white supremacy and settler colonialism. *Progress in Human Geography* 40(6): 715–33.

Bonnett, Alastair. 1997. Geography, 'race' and whiteness: invisible traditions and current challenges. *Area* 29(3): 193–9.

Bonnett, Alastair. 2014. *White Identities: An Historical and International Introduction*. London, Routledge.

Bozhkov, Elissa, Walker, C., McCourt, V. and Castleden, H. 2020. Are the natural sciences ready for truth, healing, and reconciliation with Indigenous peoples in Canada? *Journal of Environmental Studies and Sciences* 10: 226–41.

Braverman, Irus. 2021. Environmental justice, settler colonialism and more-than-humans in the occupied West Bank: an introduction. *Nature and Society* 4(1): 3–27.

Briggs, John and Sharp, J. 2004. Indigenous knowledges and development: a postcolonial caution. *Third World Quarterly* 25(4): 661–76.

Bristol Black Archives Partnership. undated. *A Guide to African-Caribbean Sources in Bristol's Museums, Galleries and Archives*. Bristol, City Council.

Brown, Jabari, Connell, K., Firth, J. and Hilton, T. 2020. The history of the land: a relational and place-based approach for teaching (more) radical food geographies. *Human Geography* 13(3): 242–52.

Brown, Leslie and Strega, S. (eds). 2015. *Research as Resistance: Revisiting Critical, Indigenous, and Anti-Oppressive Approaches*. 2nd edition. Toronto, Canadian Scholars.

Bryan, Joe and Wood, D. 2015. *Weaponizing Maps: Indigenous Peoples and Counterinsurgency in the Americas*. New York, Guildford Press.

Bryn Mawr. undated. Geology and colonialism reading list.

Bryn Mawr College, PA. At http://mineralogy.digital.brynmawr.edu/blog/geology-colonialism-reading-list.

Byrne, Jason and Wolch, J. 2009. Nature, race, and parks: past research and future directions for geographic research. *Progress in Human Geography* 33(6): 743–65.

Cameron, Emilie, de Leeuw, S. and Desbiens, C. 2014. Indigeneity and ontology. *cultural geographies* 21(1): 19–26.

Carey, David and Torres, M.G. 2010. Precursors to femicide: Guatemalan women in a vortex of violence. *Latin American Research Review* 45(3): 142–64.

Carey, Mark, Jackson, M., Antonello, A. and Rushing, J. 2016. Glaciers, gender and science: a feminist glaciology framework for global environmental change research. *Progress in Human Geography* 40(6): 770–93.

Carter, Jennifer and Hollinsworth, D. 2017. Teaching Indigenous geography in a neo-colonial world. *Journal of Geography in Higher Education* 41(2): 182–97.

Castleden, Heather, Daley, K., Morgan, V. and Sylvestre, P. 2013. Settlers unsettled: using field schools and digital stories to transform geographies of ignorance about Indigenous peoples in Canada. *Journal of Geography in Higher Education* 37(4): 487–99.

Castleden, Heather, Martin, D., Consolo, A. et al. 2017. Implementing Indigenous and western knowledge systems (Part 2): 'You have to take a backseat' and abandon the arrogance of expertise. *International Indigenous Policy Journal* 8(4), https://doi.org/10.18584/iipj.2017.8.4.8.

Castleden, Heather, Morgan, V.S. and Lamb, C. 2012a. 'I spent the first year drinking tea': exploring Canadian university researchers' perspectives on community-based participatory research involving Indigenous peoples. *The Canadian Geographer* 56(2): 160–79.

Castleden, Heather, Mulrennan, M. and Godlewska, A. 2012b. Community-based participatory research involving Indigenous peoples in Canadian geography: progress? *The Canadian Geographer* 56(2): 155–9.

Castree, N., Kitchen, R. and Rogers, A. (eds). 2013. *Oxford Dictionary of Human Geography*. Oxford, Oxford University Press.

Catungal, John P. 2019. Classroom. In Antipode Editorial Collective (ed.) *Keywords in Radical Geography*. London, Wiley, pp. 45–9.

Cavanagh, E. and L. Veracini (eds). 2016. *Handbook of the History of Settler Colonialism*. London, Routledge.

CDRE. 2013. Charter of Decolonial Research Ethics. Decoloniality Europe. At https://decoloniality europe.wixsite.com/decoloniality/ charter-of-decolonial-research-ethics.

Césaire, Aimé. 1972 [1955]. *Discourse on Colonialism*. New York, Monthly Review.

Chakrabarty, Dipesh. 2000. *Provincializing Europe: Postcolonial Thought and Historical Difference*. Princeton, Princeton University Press.

Chakrabarty, Dipesh. 2009. The climate of history: four theses. *Critical Inquiry* 35(2): 197–222.

Chantiluke, Roseanne, Kwoba, B. and Nkopo, A. 2018. Introduction from the Editors. In R. Chantiluke et al. (eds) *Rhodes Must Fall: The Struggle to Decolonise the Racist Heart of Empire*. London, Zed Books.

Chari, Sharad. 2019. Earth-Writing (Spaciousness). In Antipode Editorial Collective (ed.) *Keywords in Radical Geography*. London, Wiley, pp. 95–101.

Chatterjee, Partha. 1993. *The Nation and its Fragments*. Princeton, Princeton University Press.

Chawla, Devika and Atay, A. 2018. Introduction: decolonizing autoethnography. *Cultural Studies–Critical Methodologies* 18(1): 3–8.

CIHR, NSERC and SSHRC. 2018. *Tri-Council Policy Statement: Ethical Conduct for Research Involving Humans*. Ottawa, Institute of Health Research.

Clement, Vincent. 2017. Beyond the sham of the emancipatory Enlightenment: rethinking the relationship of Indigenous epistemologies, knowledges and geography through decolonizing paths. *Progress in Human Geography* 43(2): 276–94.

Coddington, Kate. 2017. Voice under scrutiny: feminist methods, anticolonial responses and new methodological tools. *Professional Geographer* 69(2): 314–20.

Collard, Rosemary-Claire, Dempsey, J. and Sundberg, J. 2015. A manifesto for abundant futures. *Annals of the Association of American Geographers* 105(2): 322–30.

Comaroff, Jean and Comaroff, J. 2012. *Theory from the South: Or, Why Euro-America is Evolving Toward Africa*. London, Routledge.

Connell, Raewyn. 2007. *Southern Theory*. Cambridge, Polity.

Connell, Raewyn. 2014. Using southern theory: decolonizing southern thought in theory, research and application. *Planning Theory* 13(2): 210–23.

Cook, Ian. 2000. 'Nothing can ever be the case of "us" and "them" again': exploring the politics of difference through border pedagogy and student journal writing. *Journal of Geography in Higher Education* 24(1): 13–27.

Coombes, Brad, Johnson, J.T. and Howitt, R. 2012. Indigenous geographies I: mere resource conflicts? The complexities in Indigenous land and environmental claims. *Progress in Human Geography* 36(6): 810–21.

Coombes, Brad, Johnson, J.T. and Howitt, R. 2014. Indigenous geographies III: methodological innovation and the unsettling of participatory research. *Progress in Human Geography* 38(6): 845–54.

Corntassel, Jeff. 2020. Restorying Indigenous Landscapes: Community Regeneration and Resurgence. In N.J. Turner (ed.) *Plants, People and Place*. Montreal, McGill-Queen's University Press, pp. 350–65.

Coronil, Fernando. 1996. Beyond Occidentalism: toward nonimperial geohistorical categories. *Cultural Anthropology* 11(1): 51–87.

Coronil, Fernando. 2004. Latin American Postcolonial Studies and Global Decolonization. In N. Lazarus (ed.) *Cambridge Companion to Postcolonial Literary Studies*. Cambridge, Cambridge University Press, pp. 221–40.

Coulthard, Glen S. 2014. *Red Skin, White Masks*. Minneapolis, University of Minnesota Press.

Courtheyn, Christopher. 2017. Peace geographies: expanding from modern-liberal peace to radical trans-relational peace. *Progress in Human Geography* 42(5): 741–58.

Craggs, Ruth and Neate, H. 2020. What happens if we start

from Nigeria? Diversifying histories of geography. *Annals of the Association of American Geographers* 110(3): 899–916.

Crutzen, Paul J. and Stoermer, E. 2000. The Anthropocene. *Global Change Newsletter* 41: 17–18.

Cruz, Valter. 2017. Geografia e pensamento descolonial: notas sobre um diálogo necessário para a renovação do pensamento crítico. In V. Cruz and D. de Oliveira (eds) *Geografia e Giro Descolonial*. Rio de Janeiro, Letra Capital, pp. 15–36.

Cunsolo Willox, Ashlee, Harper, S., Edge, V. and Rigolet Inuit Community Government. 2012. Storytelling in a digital age: digital storytelling as an emerging narrative method for preserving and promoting Indigenous oral wisdom. *Qualitative Research* 13(2): 127–47.

Cupples, Julie and Glynn, K. 2014. Indigenizing and decolonizing higher education in Nicaragua's Atlantic Coast. *Singapore Journal of Tropical Geography* 35(1): 56–71.

Cupples, Julie and Grosfoguel, Ramón (eds). 2018. *Unsettling Eurocentrism in the Westernized University*. London, Routledge.

Daigle, Michelle. 2016. Awawanenitakik: the spatial politics of recognition and relational geographies of Indigenous self-determination. *The Canadian Geographer* 60(2): 259–69.

Daigle, Michelle. 2019. The spectacle of reconciliation: on (the) unsettling responsibilities to Indigenous peoples in the academy. *Environment and Planning D: Society and Space* 37(4): 703–21.

Daigle, Michelle and Ramírez, M.M. 2019. Decolonial Geographies. In Antipode Editorial Collective (ed.) *Keywords in Radical Geography*. London, Wiley, pp. 78–84.

Daigle, Michelle and Sundberg, J. 2017. From where we stand: unsettling geographical knowledge in the classroom. *Transactions of the Institute of British Geographers* 42(3): 338–41.

Daley, Patricia. 2018. Reparation in the Space of the University in the Wake of Rhodes Must Fall. In R. Chantiluke et al. (eds) *Rhodes Must Fall*. London, Zed Books, pp. 72–83.

Darug Ngurra, Dadd, L., Glass, P. et al. 2019. *Yanama*

budyari gumada: reframing the urban to care as Darug Country in western Sydney. *Australian Geographer* 50(3): 279–93.

Davies, Andy. 2019. *Geographies of Anticolonialism*. London, Wiley.

Davies, Thom. 2019. Slow violence and toxic geographies: 'out of sight' for whom? *Politics and Space C*, https://doi.org/10.1177/2399654419841063.

Davis, Heather and Todd, Z. 2017. On the importance of a date, or decolonizing the Anthropocene. *ACME* 16(4): 761–80.

Davis, Rianna. 2019. Cambridge's slave trade inquiry is not enough. It must be followed by concrete action. *Huffington Post*, 1 May.

Daya, Shari. 2021. Moving from crisis to critical praxis: geography in South Africa. *Transactions of the IBG*, https://doi.org/10.1111/tran.12459.

de la Cadena, Marisol. 2010. Indigenous cosmopolitics in the Andes: conceptual reflections beyond 'politics'. *Cultural Anthropology* 25(2): 334–70.

de Leeuw, Sarah. 2013. State of care: the ontologies of child welfare in British Columbia. *cultural geographies* 21(1): 59–78.

de Leeuw, Sarah. 2016. Tender grounds: intimate visceral violence and British Colombia's colonial geographies. *Political Geography* 52: 14–23.

de Leeuw, Sarah. 2017a. Intimate colonialisms: the material and experienced places of British Columbia's residential schools. *The Canadian Geographer* 51(3): 339–59.

de Leeuw, Sarah. 2017b. Writing as righting: truth and reconciliation, poetics, and new geo-graphing in colonial Canada. *The Canadian Geographer* 61(3): 306–18.

de Leeuw, Sarah and Hunt, S. 2018. Unsettling decolonizing geographies. *Geographical Compass* 12(7): 1–14.

De Lissovoy, N. 2010. Decolonial pedagogy and the ethics of the global. *Discourse: Studies in the Cultural Politics of Education* 31(3): 279–93.

de Sousa Santos, Boaventura. 2014. *Epistemologies of the South: Justice Against Epistemicide*. London, Routledge.

de Sousa Santos, Boaventura. 2016. The University at a Crossroads. In R. Grosfoguel et al. (eds) *Decolonizing the Westernizing University*. Lexington, MD, Rowman & Littlefield, pp. 3–14.

de Sousa Santos, Boaventura. 2017. A Non-Occidentalist West? In J. Paraskeva. (ed.) *Towards a Just Curriculum Theory*. London, Routledge, pp. 67–89.

Decolonizing the Curriculum. 2017. Seminar series 'Decolonizing the Curriculum in Theory and Practice'. Centre for Research in Arts, Social Science and Humanities, University of Cambridge. Podcasts. At https://sms.cam.ac.uk/collection/2345401.

Denzin, Norman and Lincoln, Y.S. 2014. Introduction: Critical Methodologies and Indigenous Inquiry. In N. Denzin et al. (eds) *Handbook of Critical and Indigenous Methodology*. London, Sage, pp. 1–20.

Derickson, Kate. 2017. Urban geography II: urban geography in the age of Ferguson. *Progress in Human Geography* 41(2): 230–44.

Desai, Vandana. 2017. Black and Minority Ethnic (BME) student and staff in contemporary British Geography. *Area* 49(3): 320–3.

Dhillon, Carla. 2020. Indigenous feminisms: disturbing colonialism in environmental science partnerships. *Sociology of Race and Ethnicity* 6(4): 483–500.

Domosh, Mona. 2015. Why is our curriculum so white? AAG President's column. *Association of American Geographers Newsletter*, http://news.aag.org/2015/06/why-is-our-geography-curriculum-so-white.

Dorries, Heather and Ruddick, S. 2018. Between concept and context: reading Gilles Deleuze and Leanne Simpson in their in/commensurabilities. *cultural geographies* 25(4): 619–35.

Driver, Felix. 2001. *Geography Militant: Cultures of Exploration and Empire*. Oxford, Blackwell.

Duncan, James S. and Duncan, N. 2004. *Landscapes of Privilege: Aesthetics and Affluence in an American Suburb*. London, Routledge.

Dwyer, Owen and Jones, J.P. 2000. White socio-spatial

epistemology. *Social & Cultural Geography* 1(2): 209–22.

Eckstein, Lars and Schwarz, A. 2019. The making of Tupaia's map: a story of the extent and mastery of Polynesian navigation, competing systems of wayfinding on James Cook's Endeavour, and the invention of an ingenious cartographic system. *The Journal of Pacific History* 54(1): 1–95.

Eddo-Lodge, Reni. 2018. *Why I'm No Longer Speaking to White People About Race*. London, Bloomsbury.

Elden, Stuart. 2013. *The Birth of Territory*. Chicago, University of Chicago Press.

Elder, Brent and Odoyo, K. 2018. Multiple methodologies: using community-based participatory research and decolonizing methodologies in Kenya. *International Journal of Qualitative Studies* 31(4): 293–311.

Elliott-Cooper, Adam. 2017. 'Free, decolonised education': a lesson from the South African student struggle. *Area* 49(3): 332–4.

Ellis, C., Adams, T. and Bochner, A.P. 2011. Autoethnography: an overview. *Historical Social Research–Historische Sozialforschung* 36(4), Art. 10.

Escobar, Arturo. 2007. Worlds and knowledges otherwise. *Cultural Studies* 21(2–3): 179–210.

Escobar, Arturo. 2008. *Territories of Difference: Place, Movements, Life, Redes*. Durham, NC, Duke University Press.

Esson, James. 2018. 'The why and the white': racism and curriculum reform in British geography. *Area*, https://doi.org/10.1111/area.12475.

Esson, James, Noxolo, P., Baxter, R., Daley, P. and Byron, M. 2017. The 2017 RGS-IBG chair's theme: decolonising geographical knowledges, or reproducing coloniality? *Area* 49: 384–8.

Esterhuysen, Amanda, Knight, J. and Keartland, T. 2018. Mine waste: the unseen dead in a mining landscape. *Progress in Physical Geography* 42(5): 650–66.

Fanon, Frantz. 2004 [1963]. *The Wretched of the Earth*. New York, Grove Press.

Fanon, Frantz. 2008 [1952]. *Black Skin, White Masks.* London, Pluto.

Faria, Caroline, Falola, B., Henderson, J. and Torres, R. 2019. A long way to go: collective paths to racial justice in geography. *Professional Geographer* 71(2): 364–76.

Ferretti, Federico. 2019. History and philosophy of geography I: decolonising the discipline, diversifying archives and historicising radicalism. *Progress in Human Geography,* https://doi.org/10.1177/0309132519893442.

Freire, Paolo. 1971. *Pedagogy of the Oppressed.* New York, Continuum.

Galeano, Eduardo. 1971. *The Open Veins of Latin America.* New York, Monthly Review Press.

Garba, Tapji and Sorentino, S.-M. 2020. Slavery is a metaphor: a critical commentary on Eve Tuck and K. Wayne Yang's 'Decolonization is Not a Metaphor'. *Antipode* 52(3): 764–82.

Gaztambide-Fernández, Ruben. 2012. Decolonization and the pedagogy of solidarity. *Decolonization: Indigeneity, Education and Society* 1(1): 41–67.

Gentelet, Karine, Basile, S. and Asselin, H. 2018. 'We have to start sounding the trumpet for things that are working': on concrete ways to decolonize research. *ACME* 17(3): 832–39.

Gillborn, David. 2006. Critical Race Theory and education: racism and anti-racism in educational theory and practice. *Discourse: Studies in the Cultural Politics of Education* 27: 11–32.

Gill, Hartej, Purru, K. and Lin, G. 2012. In the midst of participatory action research practice: moving towards decolonizing and decolonial praxis. *Reconceptualizing Educational Research Methodology* 3(1): 1–15.

Godlewska, Anne, Moore, J. and Badnasek, C. 2010. Cultivating ignorance of Aboriginal realities. *The Canadian Geographer* 54(4): 417–40.

Goldberg, Susan. 2018. For decades, our coverage was racist. To rise above the past, we must acknowledge it. *National Geographic*, 'The Race Issue', 12 March.

Gordon, Neve and Ram, M. 2016. Ethnic cleansing and

the formation of settler colonial geographies. *Political Geography* 53: 20–9.

Gregory, Derek. 2004. *The Colonial Present*. Oxford, Blackwell.

Gregory, Derek, Johnston, R., Pratt, G., Watts, M.J. and Whatmore, S. (eds). 2009. *The Dictionary of Human Geography*. Oxford, Wiley-Blackwell.

Grosfoguel, Ramón. 2007. The epistemic decolonial turn: beyond political-economy paradigms. *Cultural Studies* 21(2–3): 211–23.

Grosfoguel, Ramón. 2013. The structure of knowledge in westernized universities: epistemic racism/sexism and the four genocides/epistemicides of the long sixteenth century. *Human Architecture* XI(1): 73–90.

Grosfoguel, Ramón. 2017. Decolonising Western Universalism: Decolonial Pluri-versalism from Aimé Césaire to the Zapatistas. In J.M. Paraskeva (ed.) *Towards a Just Curriculum Theory*. London, Routledge, pp. 147–64.

Grosfoguel, Ramón. 2019. What is Racism? Zone of Being and Zone of Non-Being. In J. Cupples and R. Grosfoguel (eds) *Unsettling Eurocentrism in the Westernized University*. London, Routledge, pp. 264–73.

Grove, A.T. 2010. National life stories: an oral history of British science. A.T. Grove. Interviewed by Paul Merchant. C1379/12. London, British Library.

Guardian. 2020. Supreme court declares large part of east Oklahoma to be Native American land. *Guardian*, 10 July.

Guthman, Julie. 2008. Bringing good food to others: investigating the subjects of alternative food practice. *cultural geographies* 15: 431–47.

Haesbaert, Rogério. 2011. *El mito de la desterritorialización: del fin de los territorios a la multiterritorialidad*. Mexico City, Siglo XXI.

Haesbaert, Rogério. 2013. A Global Sense of Place and Multiterritorialities. In D. Featherstone and J. Painter (eds) *Spatial Politics*. Oxford, Wiley-Blackwell, pp. 146–56.

Haesbaert, Rogério. 2021. *Território e Descolonialidade: Sobre o giro (multi)territorial/de(s)colonial na América*

Latina. Buenos Aires, CLACSO/Universidade Federal Fluminense.

Hale, Charles and Stephen, L. (eds). 2013. *Otros Saberes: Collaborative Research on Indigenous and Afro-Descendant Cultural Politics*. Santa Fe, SAR Press.

Halvorsen, Sam. 2018. Decolonizing territory: dialogues with Latin America knowledges and grassroots strategies. *Progress in Human Geography* 43(5): 790–814.

Haraway, Donna. 1988. Situated knowledges: the science question in feminism and the privilege of partial perspective. *Feminist Studies* 14(3): 575–99.

Harris, Jonathan. 2020. Imazighen of France: articulations of an indigenous diaspora. *Journal of Ethnic and Migration Studies*, https://doi.org/10.1080/136 9183X.2020.1788382.

Harvey, David. 1973. *Social Justice and the City*. London, Edward Arnold.

Harvey, Neil. 2016. Practicing autonomy: Zapatismo and decolonial liberation. *Latin American and Caribbean Ethnic Studies* 11(1): 1–24.

Hill Collins, Patricia. 2015. Intersectionality's definitional dilemmas. *Annual Review of Sociology* 41: 1–20.

Hinton, Mark and Ono-George, M. 2019. Teaching a history of 'race' and anti-racist action in an academic classroom. *Area* 52(4): 716–21 .

Holmes, Cindy, Hunt S. and Piedalue, A. 2014. Violence, colonialism and space: towards a decolonising dialogue. *ACME* 14(2): 539–570.

hooks, bell. 1990. Marginality as a Site of Resistance. In R. Ferguson and T. Minh-ha (eds) *Out There: Marginalization and Contemporary Cultures*. Cambridge, MA, MIT Press, pp. 241–3.

Hovorka, Alice J. 2017. Animal geographies I: globalizing and decolonizing. *Progress in Human Geography* 41(3): 382–94.

Howard-Wagner, Deirdre, Bargh, M. and Altamirano-Jimenez, I. (eds). 2018. *The Neoliberal State, Recognition and Indigenous Rights*. Acton, ANU Press.

Howitt, Richard. 2001a. Frontiers, borders, edges: liminal

challenges to the hegemony of exclusion. *Australian Geographical Studies* 39(2): 233–45.

Howitt, Richard. 2001b. Constructing engagement: geographical education for justice within and beyond tertiary classrooms. *Journal of Geography in Higher Education* 25(2): 147–66.

Howitt, Richard. 2020. Unsettling the taken (for granted). *Progress in Human Geography* 44(2): 193–215.

Hunt, Sarah. 2014. Ontologies of indigeneity: the politics of embodying a concept. *cultural geographies* 21(1): 27–32.

Hunt, Sarah and Holmes, C. 2015. Everyday decolonization: living a decolonizing queer politics. *Journal of Lesbian Studies* 19(2), 154–72.

Icaza, Rosalba and de Jong, S. 2018. Introduction. In S. de Jong et al. (eds) *Decolonization and Feminisms in Global Teaching and Development*. London, Routledge, pp. xv–xxxiv.

Inwood, Joshua and Yarborough, R.A. 2010. Racialized places, racialized bodies: the impact of racialization on individual and place identity. *Geojournal* 75: 299–301.

Ioris, Antonio, Benites, T. and Goettert, J. 2019. Challenges and contributions of indigenous geography: learning with and for the Kaiowa-Guarani of South America. *Geoforum* 102: 137–41.

Jackson, Mark. 2014. Composing postcolonial geographies: postconstructivism, ecology and overcoming ontologies of critique. *Singapore Journal of Tropical Geography* 35: 72–87.

Jazeel, Tariq. 2011. Spatializing difference beyond cosmopolitanism: rethinking planetary futures. *Theory, Culture and Society* 28(5): 75–97.

Jazeel, Tariq. 2014. Subaltern geographies: geographical knowledge and postcolonial strategy. *Singapore Journal of Tropical Geography* 35: 88–103.

Jazeel, Tariq. 2017. Mainstreaming geography's decolonising imperative. *Transactions of the Institute of British Geographers* 42: 334–7.

Jazeel, Tariq. 2019. *Postcolonialism*. London, Routledge.

Jazeel, Tariq and Legg, S. (eds). 2019. *Subaltern Geographies*. London, University of Georgia Press.

Jazeel, Tariq and McFarlane, C. 2010. The limits of responsibility: a postcolonial politics of academic knowledge production. *Transactions of the Institute of British Geographers* 35: 109–24.

Johnson, Azeezat. 2018. An Academic Witness: White Supremacy within and Beyond Academia. In A. Johnson et al. (eds) *The Fire Now: Anti-Racist Scholarship in Times of Explicit Racial Violence*. London, Zed Books, pp. 15–25.

Johnson, Azeezat. 2020a. Throwing our bodies against the white background of academia. *Area* 52(1): 89–96.

Johnson, Azeezat. 2020b. Refuting 'How the other half lives': I am a woman's rights. *Area* 52(4): 801–5.

Johnson, Jay T., Cant, G., Howitt, R. and Peters, E. 2007. Creating anti-colonial geographies: embracing Indigenous peoples' knowledges and rights. *Geographical Research* 45(2): 117–20.

Johnson, Jay T., Howitt, R. et al. 2016. Weaving Indigenous and sustainability sciences to diversify our methods. *Sustainability Science* 11: 1–11.

Johnson, Jay T. and Murton, B. 2007. Re/placing native science: Indigenous voices in contemporary constructions of nature. *Geographical Research* 45(2): 121–9.

Kamaoli-Kuwada, Bryan. 2015. We live in the future. Come join us. *Kekaupu Hehiale* Blog. At hehiale.com/2015/04/03/we-live-in-the-future-come-join-us.

Katundu, Mangasini. 2020. Which road to decolonizing the curricula? Interrogating African higher education futures. *Geoforum* 115: 150–2.

Kauffman, Craig M. and Martin, P.L. 2018. Constructing rights of nature norms in the US, Ecuador, and New Zealand. *Global Environmental Politics* 18(4): 43–62.

Kedar, Alexandre, Amara, A. and Yiftachel, O. 2018. *Emptied Lands: A Legal Geography of Bedouin Rights in the Negev*. Stanford, Stanford University Press.

Kershaw, Geoffrey, Castleden, H. and Laroque, C. 2014. An argument for ethical physical geography research on Indigenous landscapes in Canada. *The Canadian Geographer* 58(4): 393–9.

Kidman, Joanna, MacDonald, L., Funaki, H., Ormond, A., Southon, P. and Tomlins-Jahnkne, H. 2021. 'Native time' in the white city: indigenous youth temporalities in settler-colonial space. *Children's Geographies* 19(1): 24–36.

Kinsman, Phil. 1995. Landscape, race and national identity: the photography of Ingrid Pollard. *Area* 27(4): 300–10.

Knight, Jasper. 2018. Decolonizing and transforming the geography undergraduate curriculum in South Africa. *South African Geographical Journal* 100(3): 271–90.

Knitter, D., Augustin, K., Biniyaz, E. et al. 2019. Geography and the Anthropocene: critical approaches needed. *Progress in Physical Geography* 43(3): 451–61.

Kobayashi, Audrey. 1999. 'Race' and racism in the classroom: some thoughts on unexpected moments. *Journal of Geography* 98: 179–82.

Kobayashi, Audrey. 2002. The Construction of Geographical Knowledge – Racialization, Spatialization. In K. Anderson et al. (eds) *Handbook of Cultural Geography*. London, Sage, pp. 544–56.

Kobayashi, Audrey and de Leeuw, S. 2010. Colonialism and the Tensioned Landscapes of Indigeneity. In S. Smith et al. (eds) *The Sage Handbook of Social Geographies*. London, Sage, pp. 118–38.

Kobayashi, Audrey and Peake, L. 2000. 'Racism out of place': thoughts on an anti-racist agenda for geography in the new millennium. *Annals of the Association of American Geographers* 90(1): 391–403.

Koch, Alexander, Brierley, C., Maslin, M.M. and Lewis, S.L. 2019. Earth system impacts of the European arrival and Great Dying in the Americas after 1492. *Quaternary Science Reviews* 207: 13–36.

Kollectiv Orangotango+. 2018. *This is Not an Atlas: A Global Collection of Counter-Cartographies*. Bielefeld, transcript Verlag, https://www.transcript-verlag.de/shopMedia/openaccess/pdf/oa9783839445198.pdf.

Koopman, Sara. 2019. Peace. In Antipode Editorial Collective (ed.) *Keywords in Radical Geography*. London, Wiley, pp. 207–11.

Kuokkanen, Ruana. 2008. What is hospitality in the academy? Epistemic ignorance and the (im)possible gift. *Review of Education, Pedagogy and Cultural Studies* 30(1): 60–82.

Langdon, Jonathan. 2013. Decolonising development studies: reflections on critical pedagogies in action. *Canadian Journal of Development Studies* 34(3): 384–99.

LaRocco, A.A., Shinn, J.E. and Madise, K. 2019. Reflections on positionalities in social science fieldwork in Northern Botswana: a call for decolonizing research. *Politics & Gender* 16(3): 845–73.

Larsen, Soren C. and Johnson, J.T. 2012. In between worlds: place, experience, and research in Indigenous geography. *Journal of Cultural Geography* 29(1): 1–13.

Larsen, Soren C. and Johnson, J.T. 2016. The agency of place: toward a more-than-human geographical self. *GeoHumanities* 2(1): 149–66.

Lave, Rebecca. 2015. Introduction to special issue on critical physical geography. *Progress in Physical Geography* 39(5): 571–5.

Law, John. 2015. What's wrong with a one-world world? *Distinktion: Scandinavian Journal of Social Theory* 16(1): 126–39.

Lazarus, Neil (ed.). 2004. *The Cambridge Companion to Postcolonial Literary Studies*. Cambridge, Cambridge University Press.

Lee, Hyunjung and Cho, Y. 2012. Introduction: colonial modernity and beyond in East Asian contexts. *Cultural Studies* 26(5): 601–16.

Lefebvre, Henri. 1991. *The Production of Space*. Oxford, Blackwell.

Leibowitz, Brenda, Mgqwashu, E., et al. 2019. Decolonising research: the use of drawings to facilitate place-based biographic research in southern Africa. *Journal of Decolonising Disciplines* 1(1): 27–46.

Leonardo, Zeus. 2018. Dis-orienting western knowledge. *Cambridge Journal of Anthropology* 36(2): 7–20.

Livingstone, David. 1992. *The Geographical Tradition.* Oxford, Blackwell.

Loftus, Alex. 2019. Political ecology I: where is political ecology? *Progress in Human Geography* 43(1): 172–82.

Loomba, Ania. 2005. *Colonialism/Postcolonialism.* 2nd edition. London, Routledge.

Lorimer, Jamie. 2012. Multinatural geographies for the Anthropocene. *Progress in Human Geography* 36(5): 593–612.

Louis, Renee P. 2007. Can you hear us now? Voices from the margins – using Indigenous methodologies in geographical research. *Geographical Research* 45(2): 130–9.

Lucchesi, Annita H. 2019. Mapping geographies of Canadian colonial occupation: pathway analysis of murdered indigenous women and girls. *Gender, Place and Culture* 26(6): 868–87.

Lugones, Maria. 2007. Heterosexualism and the colonial/modern gender system. *Hypatia* 22(1): 186–219.

Lugones, Maria. 2010. Toward a decolonial feminism. *Hypatia* 25(4): 742–59.

Luke, Timothy W. 2018. Tracing race, ethnicity and civilization in the Anthropocene. *Society and Space* 38(1): 129–46.

McAlvay, Alex, Armstrong, C., Baker, J., Elk, L.B. and Bosco, S. 2021. Ethnobiology phase VI: decolonizing institutions, projects and scholarship. *Journal of Ethnobiology* 41(2): 170–91.

McClintock, Anne. 1992. The angel of progress: pitfalls of the term 'postcolonialism'. *Social Text* 31/32: 84–98.

MacDonald, Katherine. 2017. My experiences with Indigenist methodologies. *Geographical Research* 55(4): 369–78.

McDowell, Linda and Sharp, J.P. (eds). 1999. *A Feminist Glossary of Human Geography.* London, Arnold.

McEwan, Cheryl. 2018. *Postcolonialism, Decoloniality and Development.* 2nd edition. London, Routledge.

McGuinness, Mark. 2000. Geography matters? Whiteness and contemporary geography. *Area* 32(2): 225–30.

McIntosh, Peggy. 1989. White privilege: unpacking the invisible knapsack. National Seed Project on Inclusive Curriculum. Available at nationalseedproject.org.

McKittrick, Katherine. 2011. On plantations, prisons, and a black sense of place. *Social and Cultural Geography* 112(8): 947–63.

McKittrick, Katherine. 2013. Plantation Futures. *small axe* 17(3): 1–15.

McKittrick, Katherine. 2014. Mathematics Black Life. *The Black Scholar* 44: 16–28.

McKittrick, Katherine. 2019. Rift. In Antipode Editorial Collective (ed.) *Keywords in Radical Geography*. London, Wiley, pp. 243–7.

McKittrick, Katherine and Woods, C. 2007. *Black Geographies and the Politics of Place*. Toronto, Between the Lines.

McLean, J., Graham, M., Suchet-Pearson, S. et al. 2019. Decolonising strategies and neoliberal dilemmas in a tertiary institution: nurturing care-full approaches in a blended learning environment. *Geoforum* 101: 122–31.

Mahtani, Minelle. 2014. Toxic geographies: absences in critical race thought and practice in social and cultural geography. *Social and Cultural Geography* 15(4): 359–67.

Maldonado-Torres, Nelson. 2007. On the coloniality of being: contributions to the development of a concept. *Cultural Studies* 21(2/3): 240–70.

Maldonado-Torres, Nelson. 2011. Thinking through the decolonial turn: post-continental interventions in theory, philosophy and critique – an introduction. *Transmodernity* 2(1): 1–15.

Maldonado-Torres, Nelson. 2016. Outline of ten theses on coloniality and decoloniality. Franz Fanon Foundation. At http://franzfanonfoundation.

Mark, Glenis and Boulton, A. 2017. Indigenising photovoice: putting Māori cultural values into a research method. *Forum Qualitative Sozialforschung* 18(3): article 9.

Massey, Doreen. 2004. Geographies of responsibility. *Geografiska Annaler B: Human Geography* 86(1): 5–18.

Massey, Doreen. 2005. *For Space*. London, Sage.

Masuda, Jeffrey R., Franks, A., Kobayashi, A. and Wideman, T. 2020. After dispossession: an urban rights praxis of remaining in Vancouver's Downtown Eastside. *Society and Space* 38(2), 229–47.

Mbembe, Achille. 2010. Africa in theory: a conversation between Jean Comaroff and Achille Mbembe. *Anthropological Quarterly* 82(3): 653–78.

Mbembe, Achille. 2016. Decolonizing the university: new directions. *Arts & Humanities in Higher Education* 15(1): 29–45.

Meekosha, Helen. 2011. Decolonising disability: thinking and acting globally. *Disability and Society* 26: 667–82.

Meer, Talia and Müller, Alex. 2021. The messy work of decolonial praxis: insights from a creative collaboration among queer African youth. *Feminist Theory*, https://doi.org/10.1177/1464700121994073.

Mendoza, Breny. 2015. Coloniality of Gender and Power: From Postcoloniality to Decoloniality. In L. Disch and M. Hawkesworth (eds) *The Oxford Handbook of Feminist Theory*. Oxford, Oxford University Press, pp. 1–24.

Mignolo, Walter D. 2000. *Local Histories/Global Designs: Coloniality, Subaltern Knowledges and Border Thinking*. Princeton, University of Princeton Press.

Mignolo, Walter D. 2002. The geopolitics of knowledge and the colonial difference. *South Atlantic Quarterly* 101(1): 57–96.

Mignolo, Walter D. 2007. Delinking: the rhetoric of modernity, the logic of coloniality and the grammar of de-coloniality. *Cultural Studies* 21(2–3): 449–514.

Mignolo, Walter D. 2009. Epistemic disobedience, independent thought and de-colonial freedom. *Theory, Culture & Society* 27(7–8): 1–23.

Mignolo, Walter D. and Walsh, C.E. 2018. *On Decoloniality: Concepts, Analytics, Praxis*. Durham, NC, Duke University Press.

Miller, Theresa. 2019. *Plant Kin: A Multispecies Ethnography in Indigenous Brazil*. Austin, University of Texas Press.

Mollett, Sharlene. 2017. Irreconcilable differences? A postcolonial intersectional reading of gender, development and human rights in Latin America. *Gender, Place and Culture* 24(1): 1–17.

Mollett, Sharlene and Faria, C. 2018. The spatialities of intersectional thinking: fashioning feminist geographic futures. *Gender, Place and Culture* 25(4): 565–77.

Motta, Sara C. 2018. Feminizing and Decolonizing Higher Education: Pedagogies of Dignity in Colombia and Mexico. In S. de Jong et al. (eds) *Decolonization and Feminisms in Global Teaching and Development*. London, Routledge, pp. 25–42.

Muldoon, J. 2019. Academics: it's time to get behind decolonising the curriculum. *Guardian*, 19 March.

Munck, Ronaldo. 2020. *Social Movements in Latin America*. Newcastle-upon-Tyne, Agenda Publishing.

Mungwini, Pascah. 2013. African modernities and the critical appropriation of Indigenous knowledges: towards a polycentric global epistemology. *International Journal of African Renaissance Studies* 8(1): 78–93.

Murton, Brian. 2012. Being in the place world: toward a Māori 'geographical self'. *Journal of Cultural Geography* 29(1): 87–104.

Mutua, Kagendo and Swadener, B. (eds). 2004. *Decolonizing Research in Cross-Cultural Contexts: Critical Personal Narratives*. New York, SUNY Press.

Nakata, Martin, Nakata, V., Keech, S. and Bolt, R. 2012. Decolonial goals and pedagogies for Indigenous studies. *Decolonization: Indigeneity, Education and Society* 1(1): 120–40.

Naylor, Lindsay, Daigle, M., Zaragocín, S. and Ramírez, M.M. 2018. Interventions: bringing the decolonial to political geography. *Political Geography* 66: 199–209.

Ndlovu-Gatsheni, Sabelo. 2013. Why decoloniality in the 21st century? *The Thinker* 48: 10–15.

Neale, Timothy, Carter, R. and Nelson, T. 2019. Walking together: a decolonising experiment in bushfire management on Dja Dja Wurrung country. *cultural geographies* 26(3): 341–59.

Nhemachena, Artwell, Mlambo, N. and Kaundja, M. 2016. The notion of the 'field' and practices of researching and writing Africa: towards decolonial praxis. *Africology: Journal of Pan-African Studies* 9(7): 15–36.

Nirmal, Padini. 2016. Being and Knowing Differently in Living Worlds: Rooted Networks and Relational Webs in Indigenous Geographies. In W. Harcourt (ed.) *Palgrave Handbook of Gender and Development*. London, Palgrave Macmillan, pp. 232–50.

Norcup, Joanne. 2015a. Geography education, grey literature and the geographical canon. *Journal of Historical Geography* 49: 61–74.

Norcup, Joanne. 2015b. Awkward geographies? An historical and cultural geography of the journal *Contemporary Issues in Geography and Education* (CIGE), 1983–1991. Unpublished PhD thesis, University of Glasgow.

Noxolo, Patricia. 2017a. Decolonial theory in a time of the re-colonisation of UK research. *Transactions of the IBG* 42(3): 342–4.

Noxolo, Patricia. 2017b. Introduction: decolonising geographical knowledge in a colonised and recolonising postcolonial world. *Area* 49(3): 317–19.

Noxolo, Patricia, Raghuram, P. and Madge, C. 2012. Unsettling responsibility: postcolonial interventions. *Transactions of the Institute of British Geographers* 37(3): 418–29.

Nursey-Bray, Melissa. 2019. Uncoupling binaries, unsettling narratives and enriching pedagogical practice: lessons from a trial to Indigenize geography curricula at the University of Adelaide, Australia. *Journal of Geography in Higher Education* 43(3): 323–42.

Nxumalo, Fikile and Ross, K.M. 2019. Envisioning Black space in environmental education for young children. *Race, Ethnicity and Education* 22(4): 502–24.

Oslender, Ulrich. 2019. Geographies of the pluriverse: decolonial thinking and ontological conflict on Colombia's Pacific Coast. *Annals of the Association of American Geographers* 109(6): 1691–705.

Parnell, Susan and Oldfield, S. (eds). 2014. *Routledge Handbook on Cities of the Global South*. London, Routledge.

Peake, Linda and Kobayashi, A. 2002. Policies and practices for an antiracist geography at the millennium. *The Professional Geographer* 54: 50–61.

Pete, Shauneen. undated. 100 ways to Indigenize and decolonize academic programs and courses. University of Regina, Canada. At https://www.uregina.ca/president/assets/docs/president-docs/indigenization/indigenize-decolonize-university-courses.pdf.

Peters, Michael A. 2015. Why is my curriculum white? *Educational Philosophy and Theory*, 47(7): 641–6.

Pickerill, Jenny. 2018. Review of *Decolonising the University*, edited by G.K. Bhambra, D. Gebrial and K. Nişancıoğlu. *Times Higher Educational Supplement*, 27 September.

Pierre, Beaudelaine, Petigny, N. and Nagar, R. 2020. Embodied Translations: Decolonizing Methodologies of Knowing and Being. In A. Datta, P. Hopkins et al. (eds) *Routledge Handbook of Gender and Feminist Geographies*. London, Routledge, pp. 401–9.

Porto-Gonçalvez, C.W. 2017. De saberes e de territorios. In V. Cruz and D. de Oliveira (eds) *Geografia e giro descolonial*. Rio de Janeiro, Letra Capital, pp. 37–54.

Power, Marcus. 1998. The dissemination of development. *Environment and Planning D: Society and Space* 16(5): 577–98.

Pulido, Laura. 2000. Rethinking environmental racism: white privilege and urban development in Southern California. *Annals of the Association of American Geographers* 90(1): 12–40.

Pulido, Laura. 2002. Reflections on a white discipline. *Professional Geographer* 54(1): 42–9.

Pulido, Laura. 2017. Geographies of race and ethnicity II: environmental racism, racial capitalism and state-sanctioned violence. *Progress in Human Geography* 41(4): 524–33.

Pulido, Laura. 2018. Geographies of race and ethnicity III: settler colonialism and nonnative people of color. *Progress in Human Geography* 42(2): 309–18.

Quijano, Aníbal. 2000. Coloniality of power, Eurocentrism, and Latin America. *Nepantla: View from the South* 1(3): 533–80.

Quijano, Aníbal. 2007. Coloniality and modernity/rationality. *Cultural Studies* 21(2–3): 168–72.

Radcliffe, Sarah A. 2012. Development for a postneoliberal

era? *Sumak Kawsay*, living well and the limits to decolonization in Ecuador. *Geoforum* 43: 240–9.

Radcliffe, Sarah A. 2015. *Dilemmas of Difference: Indigenous Women and the Limits of Postcolonial Development Policy*. Durham, NC, Duke University Press.

Radcliffe, Sarah A. 2017a. Decolonising geographical knowledges. *Transactions of the Institute of British Geographers* 42(3): 329–33.

Radcliffe, Sarah A. 2017b. Geography and indigeneity I: indigeneity, coloniality and knowledge. *Progress in Human Geography* 41(2): 220–9.

Radcliffe, Sarah A. 2019. Pachamama, Subaltern Geographies, and Decolonial Projects in Andean Ecuador. In T. Jazeel and S. Legg (eds) *Subaltern Geographies*. Athens, University of Georgia Press, pp. 119–41.

Radcliffe, Sarah A. 2020. Geography and indigeneity III: co-articulation of colonialism and capitalism in indigeneity's economies. *Progress in Human Geography* 44(2): 374–88.

Radcliffe, Sarah A. and Radhuber, I.M. 2020. The political geographies of D/decolonization: variegation and decolonial challenges of/in geography. *Political Geography* 78, https://doi.org/10.1016/j.polgeo.2019.102128.

Raghuram, Parvati, Madge, C. and Noxolo, P. 2009. Rethinking responsibility and care for a postcolonial world. *Geoforum* 40(1): 5–13.

Ramírez, Margaret M. 2015. The elusive inclusive: Black food geographies and racialized food spaces. *Antipode* 47(3): 748–69.

RETOS. 2018. About the Transnational Network Other Knowledges: La Red Transnacional Otros Saberes (RETOS): Between Crisis and Other Possible Worlds. In S. de Jong et al. (eds) *Decolonization and Feminisms in Global Teaching*. London, Routledge, pp. 128–36.

Rice, Alan. 2012. The History of the Transatlantic Slave Trade and Heritage from Below: Guerrilla Memorialisation in the Era of Bicentennial Commemoration. In I. Richardson (ed.) *Heritage from Below*. London, Routledge, pp. 209–35.

Riddell, Julia K., Salamanca, A., Pepler, D.J., Cardinal, S. and McIvor, O. 2017. Laying the groundwork: a practical guide for ethical research with Indigenous communities. *International Indigenous Policy Journal* 8(2): 1–22.

Rivera Cusicanqui, Silvia. 2012. Ch'ixinakax utxiwa: a reflection on the practices and discourses of decolonization. *South Atlantic Quarterly* 111(1): 95–109.

Robinson, Jennifer and Roy, A. 2016. Debate on global urbanism and the nature of urban theory. *International Journal of Urban and Regional Research* 40(1): 181–6.

Roy, Ananya. 2016. Divesting from whiteness: the university in the age of Trumpism. *Society + Space*. At https://www.societyandspace.org/articles/divesting-from-whiteness-the-university-in-the-age-of-trumpism.

Rutazibwa, Olivia. 2018. On Babies and Bathwater: Decolonizing International Development Studies. In S. de Jong et al. (eds) *Decolonization and Feminisms in Global Teaching and Development*. London, Routledge, pp. 158–80.

Said, Edward. 1978. *Orientalism*. New York, Vintage.

Santos, Milton. 1977. Society and space: social formation as theory and method. *Antipode* 9(1): 3–13.

Santos, Milton. 1995. World Time and World Space or Just Hegemonic Time and Space? In G. Benko and U. Strohmayer (eds) *Geography, History and Social Sciences*. Dordrecht, Kluwer, pp. 45–9.

Scott Lewis, Jovan. 2018. Releasing a tradition: diasporic epistemology and the decolonized curriculum. *Cambridge Journal of Anthropology* 36(2): 21–33.

Scott Lewis, Jovan. 2020. Subject to labor: racial capitalism and ontology in the post-emancipation Caribbean. *Geoforum*, https://doi.org/10.1016/j.geoforum.2020.06.007.

Senier, Siobhan. 2018. Where a bird's eye view shows more concrete: mapping Indigenous LA for tribal visibility and reclamation. *American Quarterly* 70(4): 941–8.

Sharp, Joanne. 2009. *Geographies of Postcolonialism*. London, Sage.

Shaw, Wendy, Herman, R.D.K. and Dobbs, G.R. 2006. Encountering indigeneity: re-imagining and decolonizing geography. *Geografiska Annaler B* 88(3): 267–76.

Shilliam, Robbie. 2015. *The Black Pacific: Anti-Colonial Struggles and Oceanic Connections*. London, Bloomsbury.

Sidaway, James. 1997. The (re)making of the western 'geographical tradition': some missing links. *Area* 29(1): 72–80.

Sidaway, James. 2000. Postcolonial geographies: an exploratory essay. *Progress in Human Geography* 24(4): 591–612.

Sidaway, James, Woon, C.Y. and Jacobs, J.M. 2014. Planetary postcolonialism. *Singapore Journal of Tropical Geography* 35: 4–21.

Sihlongonyane, Mfaniseni Fana. 2015. The challenges of theorising about the Global South: a view from an African perspective. *Africa Insight* 45(2): 59–74.

Simpson, Audra. 2007. On ethnographic refusal: indigeneity, 'voice' and colonial citizenship. *Junctures* 9: 67–80.

Simpson, Audra. 2014. *Mohawk Interruptus: Political Life Across the Borders of Settler States*. Durham, NC, Duke University Press.

Simpson, Leanne B. 2011. *Dancing on Our Turtle's Back: Stories of Nishnaabeg Re-creation, Resurgence and a New Emergence*. Winnipeg, Arbeiter Ring.

Simpson, Leanne B. 2014. Land as pedagogy: Nishnaabeg intelligence and rebellious transformation. *Decolonization: Indigeneity, Education and Society* 3(3): 1–25.

Sium, Aman, Desai, C. and Ritskes, E. 2012. Towards the 'tangible unknown': decolonization and the indigenous future. *Decolonization: Indigeneity, Education and Society* 1(1): i–xiii.

Slater, David. 1992. On the borders of social theory: learning from other regions. *Environment and Planning D: Society and Space* 10: 307–27.

Slater, David. 2004. *Geopolitics and the Postcolonial: Rethinking North–South Relations*. Oxford, Blackwell.

Slaymaker, Olav, Mulrennan, M. and Catto, N. 2020. Implications of the Anthropocene Epoch in Geomorphology. In O. Slaymaker and N. Catto (eds) *Landscapes and Landforms of Eastern Canada*, Berlin, Springer, pp. 583–8.

Smithers Graeme, Cindy and Mandawe, E. 2017. Indigenous

geographies: research as reconciliation. *International Indigenous Policy Journal* 8(2): article 2.

Smith, Neil. 2008. *Uneven Development*. 3rd edition. Athens, University of Georgia Press.

Snow, Kathy. 2018. What does being a settler ally mean? A graduate student's experience learning from and working within Indigenous research paradigms. *International Journal of Qualitative Methods* 17: 1–11.

Soja, Edward W. 1989. *Thirdspace: Journeys to Los Angeles and Other Real-and-Imagined Places*. Oxford, Blackwell.

Spivak, Gayatri C. 1988. Can the Subaltern Speak? In C. Nelson and L. Grossberg (eds) *Marxism and the Interpretation of Culture*. Basingstoke, Macmillan, pp. 271–313.

Spivak, Gayatri C. 1999. *A Critique of Postcolonial Reason*. Cambridge, MA, Harvard University Press.

Stea, David and Wisner, B. 1984. The fourth world. *Antipode* 16(2): 3–12.

Stoler, Ann L. 2008. Imperial debris: reflections on ruin and ruination. *Cultural Anthropology* 23(2): 191–219.

Stoler, Ann L. 2016. *Duress: Imperial Durabilities for Our Times*. Durham, NC, Duke University Press.

Sundaresan, Jayaraj. 2020. Decolonial reflections on urban pedagogy in India. *Area* 52(4): 722–30.

Sundberg, Juanita. 2014. Decolonizing posthumanist geographies. *cultural geographies* 21(1): 33–47.

Sundberg, Juanita. 2015. Ethics, Entanglement and Political Ecology. In T. Perreault, G. Bridge and J. McCarthy (eds) *Routledge Handbook of Political Ecology*. New York, Routledge, pp. 117–26.

Sylvestre, Paul, Castleden, H., Martin, D. and McNally, M. 2018. 'Thank you very much … you can leave our community now': geographies of responsibility, relational ethics, acts of refusal, and the conflicting requirements of academic localities in Indigenous research. *ACME* 17(3): 750–79.

Tate, Shirley A. and Bagguley, P. 2019. Introduction. In S.A. Tate and P. Bagguley (eds) *Building the Anti-Racist University*. London, Routledge, pp. 1–10.

Thomas, Amanda. 2015. Indigenous more-than-humanisms: relational ethics with the Hurunui River in Aotearoa New Zealand. *Social & Cultural Geography* 16(8): 974–90.

Tipa, Gail, Panelli, R. and the Moeraki Stream Team. 2009. Beyond 'someone else's agenda': an example of indigenous/academic research collaboration. *New Zealand Geographer* 65: 95–106.

Todd, Zoe. 2016. An Indigenous feminist's take on the ontological turn: 'ontology' is just another word for colonialism. *Journal of Historical Sociology* 29(1): 4–22.

Tolia-Kelley, Divya. 2017. A day in the life of a geographer: 'lone', black, female. *Area* 49(3): 324–8.

Tomlinson, Barbara. 2018. *Undermining Intersectionality*. Philadelphia, Temple University Press.

Townsend, Janet G. and collaborators U. Arrevillaga, J. Bain, D. Cancino, S. Frenk, S. Pancheco and E. Perez. 1994. *Women's Voices from the Rainforest*. London, Routledge.

Tuck, Eve, McKenzie, M. and McCoy, K. 2014. Land education: Indigenous, post-colonial and decolonizing perspectives on place and environmental education research. *Environmental Education Research* 20(1): 1–23.

Tuck, Eve and Yang, K.W. 2012. Decolonization is not a metaphor. *Decolonization: Indigeneity, Education and Society* 1(1): 1–40.

Tuck, Eve and Yang, K.W. 2014. R-Words: Refusing Research. In D. Paris and M.T. Winn (eds) *Humanizing Research: Decolonizing Qualitative Inquiry with Youth and Communities*. London, Sage, pp. 223–48.

Tuhiwai Smith, Linda. 2006. Colonizing Knowledges. In R. Maaka and C. Anderson (eds) *The Indigenous Experience*. Toronto, Canadian Scholars, pp. 91–113.

Tuhiwai Smith, Linda. 2012. *Decolonizing Methodologies: Research and Indigenous Peoples*. 2nd edition. London, Zed Books.

Tuhiwai Smith, Linda. 2020. Decolonising Methodologies, 20 years on: an interview with Professor Linda Tuhiwai Smith. 11 February. Goldsmiths, University of London.

Tuhiwai Smith, Linda, Tuck, E. and Yang, K.W. (eds).

2018. *Indigenous and Decolonizing Studies in Education: Mapping the Long View*. London, Routledge.

Turner, Nancy J. (ed.) 2020. *Plants, People and Places: The Roles of Ethnobotany and Ethnoecology in Indigenous Peoples' Land Rights*. Montreal, McGill-Queen's University Press.

UCL Collective. 2015. 8 reasons the curriculum is white. At https://novaramedia.com/2015/03/23/8-reasons-the-curriculum-is-white.

Underhill-Sem, Yvonne. 2020. The audacity of the ocean: gendered politics of positionality in the Pacific. *Singapore Journal of Tropical Geography* 41(3): 314–28.

Vaeau, Tarapuhi and Trundle, C. 2020. Decolonising Māori-Pākehā Research Collaborations: Towards an Ethics of Whanaungatanga and Manaakitanga in Cross-Cultural Research Relationships. In L. George et al. (eds) *Indigenous Research Ethics*. Bingley, Emerald Publishing, pp. 207–21.

Valentine, Gill. 2002. People Like Us: Negotiating Sameness and Difference in the Research Process. In P. Moss (ed.) *Feminist Geography in Practice: Research and Methods*. Oxford, Blackwell.

Van Sant, Levi, Hennessy, E. and Domosh, M. 2020. Historical geographies of, and for, the present. *Progress in Human Geography* 44(1): 168–88.

Vela-Almeida, Diana, Zaragocín, S., Bayón, M., Arrazola, I. and Mason, Dees, L. 2020. Imagining plural territories of life: a feminist reading of resistance in the socio-territorial movements in Ecuador. *Journal of Latin American Geography* 19(2): 265–87.

Velednitsky, Stepha, Hughes, S.N.S. and Machold, R. 2020. Political geographical perspectives on settler colonialism. *Geography Compass*, https://doi.org/10.1111/gec3.12490.

wa Thiong'o, Ngũgĩ. 1986. *Decolonizing the Mind: The Politics of Language in African Literature*. London, James Currey.

wa Thiong'o, Ngũgĩ. 1995 [1972]. On the Abolition of the English Department. In B. Ashcroft, G. Griffiths and H. Tiffin (eds) *The Post-colonial Studies Reader*. London, Routledge, pp. 438–41.

Wahlquist, Calla. 2016. 'It's the same story': how Australia and Canada are twinning on bad outcomes for Indigenous people. *Guardian*, 24 February.

Wainwright, Joel. 2010. On Gramsci's 'conceptions of the world'. *Transactions of the Institute of British Geographers* 35(4): 507–21.

Waldmüller, Johannes and Rodriguez, L. 2019. Buen Vivir and the rights of nature. In J. Drydyk and L. Keleher (eds) *Routledge Handbook of Development Ethics*. Abingdon, Routledge, pp. 234–47.

Walia, Harsha. 2012. Decolonizing together: moving beyond a politics of solidarity. *Briarpatch Magazine*, 1 January.

Warren, Saskia. 2021. Pluralising (im)mobililties: anti-Muslim acts and the epistemic politics of mobile methods. *Mobilities*, https://doi.org/10.1080/17450101.2021.1922068.

Watts, Michael. 2009. Decolonization. In D. Gregory et al. (eds) *Dictionary of Human Geography*. Oxford, Blackwell, pp. 145–7.

Whatmore, Sarah. 2006. Materialist returns: practising cultural geography in and for a more-than-human world. *cultural geographies* 13: 600–9.

Whyte, Kyle P. 2018. White allies, let's be honest about decolonization. *Yes Magazine*, 3 April. At https://www.yesmagazine.org/issue/decolonize/2018/04/03/white-allies-lets-be-honest-about-decolonization.

Wilcock, Deidre, Brierley, G. and Howitt, R. 2013. Ethnogeomorphology. *Progress in Physical Geography* 37(5): 573–600.

Wildcat, Matthew, McDonald, M., Irlbacher-Fox, S. and Coulthard, G. 2014. Learning from the land: Indigenous land based pedagogy and decolonization. *Decolonization* 3(3): 1–15.

Wolfe, Patrick. 2006. Settler colonialism and the elimination of the native. *Journal of Genocide Research* 8(4): 387–409.

Woodward, Emma and McTaggart, P.M. 2016. Transforming cross-cultural water research through trust, participation and place. *Geographical Research* 54(2): 129–42.

Wright, Melissa W. 2019. Border thinking, borderland

diversity and Trump's Wall. *Annals of the Association of American Geographers* 109(2): 511–19.

Wright, Sarah, Daley, L. and Curtis, F. 2021. Weathering Colonisation: Aboriginal Resistance and Survivance in the Siting of the Capital. In Kaya Barry, M. Borovnik and T. Edensor (eds) *Weather: Spaces, Mobilities, Affects*. Abingdon, Routledge, pp. 207–21.

Wright, Sarah, Lloyd, K., Suchet-Pearson, S., Burarrwanga, L., Tofa, M. and Bawaka Country. 2012. Telling stories in, through and with Country: engaging with Indigenous and more-than-human methodologies at Bawaka, NE Australia. *Journal of Cultural Geography* 29(1): 39–60.

Wright, Willie Jamal. 2020. The morphology of marronage. *Annals of the Association of American Geographers* 110(4): 1134–49.

Wylie, John. 2005. A single day's walking: narrating self and landscape on the South West Coast Path. *Transactions of the Institute of British Geographers* 30(2): 234–47.

Wynter, Sylvia. 2003. Unsettling the coloniality of being/power/truth/freedom: towards the human, after man, its overrepresentation – an argument. *CR: The New Centennial Review* 3(3): 257–337.

Ybarra, Megan. 2019. On becoming a Latinx geographies killjoy. *Society + Space*. At http://societyandspace. org/2019/01/23/on-becoming-a-latinx-geographies-killjoy.

YouGov. 2016. Rhodes must not fall. 18 January. At www. yougov.co.uk.

YouGov. 2020. How unique are British attitudes to empire? 11 March. At www.yougov.co.uk

Young, Robert J.C. 2001. *Postcolonialism: An Historical Introduction*. Oxford, Blackwell.

Young, Robert J.C. 2003. *Postcolonialism: A Very Short Introduction*. Oxford, Oxford University Press.

Yusoff, Kathryn. 2018. *A Billion Black Anthropocenes or None*. Minneapolis, University of Minnesota Press.

Yuval-Davis, Nira, Wemyss, G. and Cassidy, K. 2017. Introduction to special issue: racialized bordering discourses on European Roma. *Ethnic and Racial Studies* 40(7): 1047–57.

Yuval-Davis, Nira, Wemyss, G. and Cassidy, K. 2019. *Bordering*. Cambridge, Polity.

Zahara, Alex. 2016. Ethnographic refusal: a how to guide. *Discard Studies*, 8 August. At https://discardstudies.com/2016/08/08/ethnographic-refusal-a-how-to-guide.

Zaragocín, Sofia. 2019. Gendered geographies and elimination: decolonizing feminist geography in Latin American settler contexts. *Antipode* 51(1): 372–92.

Zaragocín, Sofia and Caretta, M. 2020. *Cuerpo-territorio*: a decolonial feminist geographical method for the study of embodiment. *Annals of the Association of American Geographers* 111(5): 1503–18.

Zurba, Melanie, Maclean, K., Woodward, E. and Islam, D. 2018. Amplifying Indigenous community participation in place-based research through boundary work. *Progress in Human Geography* 43(6): 1020–43.

Index